Nuclear Magnetic Resonance as a Tool for On-Line Catalytic Reaction Monitoring

Von der Fakultät für Mathematik, Informatik und Naturwissenschaften der Rheinisch-Westfälischen Technischen Hochschule Aachen zur Erlangung des akademischen Grades eines Doktors der Naturwissenschaften genehmigte Dissertation

von

Licenciado en Física
LISANDRO BULJUBASICH GENTILETTI

aus Santa Fé, Argentina

Berichter: Universitätsprofessor Dr. Dr. h.c. (RO) Bernhard Blümich
Universitätsprofessor Dr. Siegfried Stapf

Tag der mündlichen Prüfung: 15. März 2010

Diese Dissertation ist auf den Internetseiten der Hochschulbibliothek online verfügbar.

Berichte aus der Physik

Lisandro Buljubasich Gentiletti

Nuclear Magnetic Resonance as a Tool for On-Line Catalytic Reaction Monitoring

Shaker Verlag
Aachen 2010

Bibliographic information published by the Deutsche Nationalbibliothek
The Deutsche Nationalbibliothek lists this publication in the Deutsche Nationalbibliografie; detailed bibliographic data are available in the Internet at http://dnb.d-nb.de.

Zugl.: D 82 (Diss. RWTH Aachen University, 2010)

Copyright Shaker Verlag 2010
All rights reserved. No part of this publication may be reproduced, stored in a retrieval system, or transmitted, in any form or by any means, electronic, mechanical, photocopying, recording or otherwise, without the prior permission of the publishers.

Printed in Germany.

ISBN 978-3-8322-9126-6
ISSN 0945-0963

Shaker Verlag GmbH • P.O. BOX 101818 • D-52018 Aachen
Phone: 0049/2407/9596-0 • Telefax: 0049/2407/9596-9
Internet: www.shaker.de • e-mail: info@shaker.de

To my mother...

Contents

1 **Introduction**   1

2 **Mass Transport during H_2O_2 Decomposition**   7
   2.1 Introduction 7
   2.2 Diffusion and Flow in NMR 8
     2.2.1 General 8
     2.2.2 Free Diffusion 10
     2.2.3 Velocity Correlation and Self-Diffusion Tensor 12
     2.2.4 Pulsed Gradient Spin-Echo NMR 13
     2.2.5 Stationary and Time Dependent Random Flow 18
   2.3 Reaction Monitoring 23
     2.3.1 Brief Description 23
     2.3.2 Average Propagators During the Reaction 25
     2.3.3 D_{eff}^z and D_{eff}^x vs. Reaction Time During the Decomposition with Cu-Pellet 27
     2.3.4 D_{eff}^z vs. Δ during the Decomposition with Cu-pellet 30
     2.3.5 Comparison Between Two Different Catalyst Samples 33

3 **Chemical Exchange and Relaxation in H_2O_2**   37
   3.1 Chemical Exchange in NMR 37
     3.1.1 General 37
     3.1.2 Theoretical Treatment of Two-Site Chemical Exchange 40
   3.2 Chemical Exchange in Aqueous Hydrogen Peroxide Solutions 51
     3.2.1 Previous Works 51
     3.2.2 T_2 vs. echo time in Bulk Samples 52
   3.3 Monitoring a Catalyzed H_2O_2 Decomposition 54
     3.3.1 Monitoring the Decomposition by means of T_2 Measurements . 54

       3.3.2   The pH as an Independent Quantity 64
       3.3.3   Monitoring the Reaction by means of Simultaneous pH and
                   T_2 Measurements 70
  3.4 Exchange Rates 74

4 H_2O_2 Reaction Evolution Studied by NMR Imaging   81
  4.1 Introduction 81
  4.2 NMR Imaging 83
       4.2.1   Basics 83
       4.2.2   Spin Echo Pulse Sequence 84
       4.2.3   Two-Dimensional Imaging 86
       4.2.4   Sampling Requirements of k–space and FOV 88
       4.2.5   Contrasts and Pixel Intensity 89
  4.3 Reaction Evolution Inside the Pd-Catalyst 91
       4.3.1   The Sample 91
       4.3.2   Initial and Final States of the Reaction 92
       4.3.3   The Optimum Contrast 100
       4.3.4   Following a Reaction 103
       4.3.5   An Open Problem 107

5 Conclusions and Outlook   113

Chapter 1

Introduction

A journey of a thousand miles must begin with a single step.

Lao Zu

Traditionally, Nuclear Magnetic Resonance (NMR) used to be split according to several categories. For example, according to their application (theoretical-chemical-physical-biomedical), to the materials investigated (gas-solution-solid state-soft matter), to the techniques involved (spectroscopy-solid state techniques-imaging). Frequently, the sample determines the method, and thus the hardware required for performing the measurements. But the power of NMR, which has been declared dead for a dozen of times and is still livelier than ever, lies in its ability to combine and extend the available techniques for a more thorough solution of problems which cannot be assigned to one of the popular categories [Sta].

Thus, NMR has become a well-established method in many different areas of research. The scope of the disciplines involved is extremely broad and is still expanding, encompassing chemical, petrochemical, biological and medical research, plant physiology, aerospace engineering, process engineering, industrial food processing, materials and polymer sciences.

The work presented here must be globally seen as a combination of different aspects of Nuclear Magnetic Resonance, focused on providing a reliable tool for the optimization of chemical processes.

In the world of chemical engineering, every chemical process is designed to produce economically a desired product from a variety of starting materials through a succession of treatment steps. Figure 1.1 shows a typical situation. The raw materials undergo a number of physical treatment steps to put them in the form in which they can be reacted chemically. The products of the reaction must then undergo further physical treatment -separations, purifications, etc.- for the final desired product to be obtained. Frequently, the chemical treatment step (typically a reaction or a series of reactions taking place inside a reactor) is the hearth of the process, that makes or breaks the process economically [Lev1].

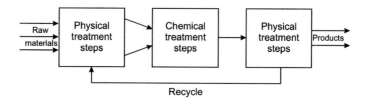

Figure 1.1: Typical chemical process.

There are many ways of classifying chemical reactions. In *chemical reaction engineering* probably the most useful scheme is the breakdown according to the number and types of phases involved, the big division being between homogeneous and heterogeneous. A reaction is homogeneous if it takes place only in one phase. A reaction is heterogeneous if it requires the presence of at least two phases to proceed at the rate that it does.

Cutting across this classification is the catalytic reaction whose rate is altered by materials, that are neither reactants nor products. These foreign materials, called **catalysts**, act as either hindering or accelerating the reaction process (they can speed the reaction by a factor of a million or much more, or they may slow the reaction) while being modified relatively slowly, if at all.

Along this work, we will be solely dealing with heterogeneous catalytic reactions, where the catalysts consists of a porous medium particle. Since in that type of reactions the solid catalyst surface is responsible for catalytic activity, a large readily accessible surface in easily handled materials is desirable. By a variety of methods, active surface areas of the order of 1000 m^2 per gram of catalyst can be obtained.

These reactions play an important role in many industrial processes, such as the production of methanol, sulfuric acid, ammonia, and various petrochemicals, polymers, paints and plastics. It is estimated that well over 50 % of all the chemicals produced today are made with the use of catalysts [Lev1].

The rate constants of the heterogeneous catalytic reactions, and therefore the efficiency of the reaction, depends on the local environment of the catalyst surface including concentration and distribution of reagents and of possible deactivating substances in the vicinity of the active sites, as well as on the rate of molecular transport as influenced by the topology and pore space geometry [RSB]. Therefore, the development and optimization of the catalyst becomes an important part of the design of a chemical process, and much effort and money are invested in that direction. For that purpose, sufficiently flexible testing methods to totally or partially quantify the results of the variants introduced in the design of different catalyst are of invaluable help. Among the methods that can be used for testing catalysts, NMR is perhaps the ultimate technique because it provides a rich toolbox for the investigation of properties on all length scales of interest while remaining strictly non-invasive. The application of the technique to small-scale reactors is of particular interest for the catalyst developers, because it makes the design-testing iteration relatively easy to perform, without much hardware requirement.

From the NMR point of view, the motivation to apply the technique to small-scale reaction units is two-fold. First, the investigation of full-scale chemical reactors used in production may be impossible due to the size or cost restriction of the hardware, or in case of NMR Imaging, may be feasible but only allow for an insufficient spatial resolution of the system, typically being of the order of one-hundredth of the resonator dimension. On the other hand, in order to follow reactions at the level of the actual reaction sites, however, studies of a single catalyst pellet at well-defined conditions can be performed with a much higher spatial resolution, allowing the verification and discrimination of coupled diffusion/reaction models. The second reason is more hardware-related, and exploits the superior performance of gradient and radiofrequency detectors on small scales, leading to the improved spatial and temporal resolution that are required to understand processes which are intrinsically fast or are localized to the submillimeter scale, such as transport dominated by self-diffusion [ABBS].

In order to show the feasibility of using parameters easily accessible experimentally with NMR to monitor heterogeneously catalyzed reactions, and then to test different aspects of the catalyst particles, a model reaction was needed. It has been opted for the following reaction:

$$2H_2O_2 \text{ (liquid)} \longrightarrow 2H_2O + O_2\text{(gas)} \quad (1.1)$$

The election of the decomposition of **aqueous hydrogen peroxide solutions** relies on two main reasons. Firstly, the fact that the reaction can be carried out in a simple laboratory glass tube, under room temperature and atmospheric pressure with occurrence of the gas phase in form of bubbles, makes it an excellent example of a simple liquid-gas reaction. It has been used recently for demonstrative purposes [Dat4, BDJ+].

Secondly, hydrogen peroxide has itself a huge importance as a chemical compound. It is one of the most versatile and environmentally desirable chemicals available today: it is used in a wide variety of industrial applications. Its oxidation capability enables hydrogen peroxide to be employed as a reactant in chemical synthesis and as a bleaching agent for paper and textiles [JC], as well as in the treatment of pollutants such as iron, sulfides, organic solvents, gasolines and pesticides. A number of biological processes do exist that produce or consume hydrogen peroxide, and it is frequently used for eliminating organic and inorganic contaminants in many environmental applications, wastewater treatment being one of the most important ones [CHL+, PBS]. The more difficult-to-oxidize pollutants may require the H_2O_2 to be activated with catalysts such as iron, copper, manganese, or other transition metal compounds. These catalysts are also used, for example, in the synthesis of conjugated polymers [SBS+], or to speed up H_2O_2 reactions that may otherwise take hours or days to complete. A totally different application is the use as a green propellant for space propulsion [Gre].

Due to the large number of applications involving hydrogen peroxide decomposition either due to equilibrium reaction processes, or supported by catalysts, there is increasing interest in the development of techniques which permit monitoring those reactions. Many analytical methods were tested for the quantification of hydrogen peroxide concentrations, including titration with potassium permanganate [HK], infrared and Raman spectroscopy [VRJM, VDS+], and ^{1}H NMR spectroscopy [SB]. For a detailed list of techniques with their respective advantages and disadvantages, see [SB] and references therein.

This thesis is organized as follows:

In **chapter 2** the feasibility of using the time dependence of the effective diffusion coefficient in the vicinity of the catalysts (or in a closed volume containing it) to monitor the decomposition with relatively high temporal resolution for several hours, is proved.

In **chapter 3** the influence of two-site chemical exchange between protons in water and hydrogen peroxide on the transverse relaxation time, and its concentration dependence in bulk samples is pointed out. It is demonstrated that for the system under investigation, the sample peculiarities, such as shape, metal content, internal magnetic field gradients and susceptibility effects, are minor compared to the dominating contrast produced by chemical exchange, thus rendering T_2 a suitable indicator of the reaction progress. In addition, it is proven that a simultaneous monitoring of transverse relaxation time and pH allows a reliable quantification of the H_2O_2 concentration at any time during the reaction. The possibility of quantifying low concentrations, under 0.1 % v/v is highlighted.

In **chapter 4** a collection of experiments is presented, in order to show how the effect of chemical exchange in the transverse relaxation time can be used as a contrast in NMR Imaging, providing a tool for monitoring the reaction with spatial resolution inside a porous particle.

In **chapter 5** a summary of the more relevant results is presented along with the respective conclusions and possible extensions of the results, and new experiments are suggested based on the obtained results.

Chapter 2

Mass Transport during H_2O_2 Decomposition

In plain words, Chaos was the law of nature, Order was the dream of man.

Henry B. Adams.

2.1 Introduction

Heterogeneously catalyzed reactions mostly take place in the presence of finely dispersed catalysts (metals such as Ni, Pt, Cu, Pd, etc.); these in turn are localized in materials of large internal surfaces, that is, porous media. The reaction efficiency then depends on parameters such as internal surface area, the homogeneity of metal distribution, and the porosity and tortuosity of the pellet. In general, the pore space of catalyst pellets is described by a complex topology, having pores in the nanometer and micrometer ranges. It is also known that the presence of micrometer-scale pores has a strong influence on reaction efficiency [Dat2, Dat1], since without these the reaction would predominantly take place at the outer edge of the pellet, and the core would remain mostly inactive. This, however, would require the use of smaller catalysts pellets for maximum efficiency, which in turn enhances pressure drop inside the fixed-bed reactor. A proper understanding of the processes governing mass transport to and from the pellet interior is therefore vital for an optimum design of the reactor.

In most reactions of technical interest, gas is one of the involved components. The gas generated during the reaction - predominantly in the vicinity of the metal sites at the pore surface - is first dissolved within the surrounding liquid phase, until the maximum solubility is exceeded. The formation of a gas phase, however, depends on the interphase tension and the size and tortuosity of the pore system; bubbles might therefore be generated inside large pores, or might only form at the external surface of the pellet. In general, each pore generates bubbles at a certain rate or frequency. Large pores lead to large bubbles at low frequency, and *vice versa*. For a constant reaction rate, these properties can be predicted for isolated pores [Dat3]. In a real catalyst pellet, however, the coupling of all the pores within the interconnected pore space gives rise to a pattern of bubble generation that cannot be computed analytically [Dat4].

Bubble formation greatly enhances not only gas transport, but also fluid transport; this affects the fluid in the vicinity of the pellets. In consequence, material transport becomes much faster than by assuming purely diffusive processes. Hence, reaction efficiency can increase dramatically.

In this chapter, the reaction H_2O_2 (liquid) $\longrightarrow$ H_2O (liquid) + 1/2 O_2 (gas) is studied in the presence of two different catalyst particles. The principal motivation was to employ NMR techniques to monitor the reaction evolution by means of the *effective diffusion coefficient* (see below) of the liquid in the vicinity of the pellet, or in a defined closed volume.

2.2 Diffusion and Flow in NMR

2.2.1 General

Self-diffusion is the random translational motion of molecules (or ions) driven by kinetic energy. Translational diffusion is the most fundamental form of transport and is a precondition for all chemical reactions, since the reacting species must collide before they can react. On the other hand, liquid flow is a phenomenon which exists in many situations in different forms. The flow can be orderly or chaotic, fast or slow, steady or time-dependent, or can be of homogeneous or inhomogeneous material and all of these differences affect the ease of measurement as well as the best method to be employed.

Flow and motion effects on an NMR signal have been known for a long time. Bloembergen et al. discussed the effect of diffusion on the NMR signal and relaxation times [BP]. The use of spin echoes to measure molecular motion was first suggested

2.2. Diffusion and Flow in NMR

by Hahn in his paper of 1950 [Hah]. Carr and Purcell measured diffusion constants in liquids by using multiple echoes and they discussed the different effects of the *first* and the *second echoes* for rephasing of flowing spins [CP]. The use of pulsed magnetic field gradients in conjunction with echoes was introduced, 11 years later, by Stejskal and Tanner in 1965 [ST].

The more standard methods for flow and diffusion measurements include scattering experiments such as X-ray, optical and ultrasound as well as more invasive experiments such as hot wire and other visualization experiments [CF]. The major attraction of NMR is that it is non-invasive, i.e. no direct contrast is necessary. One special property of NMR is that there are no preferred directions, unlike scattering experiments, which require a beam to be directed at the sample. This means that projections in any direction or orientation are, from the technological point of view, equivalents for NMR. Another difference between NMR and scattering experiments is that the typical fluid flow sample does not attenuate the r.f. signal required in the experiment. Therefore, NMR is less likely to suffer from shadows and other opacity effects. Such situations might include the study of liquids involving concentrated solids suspensions, where the solids can block the "beam"; the study of hot, cold or corrosive liquids, which require containment by materials that attenuate light and sound; or the case of studying transport properties in small-scale reactors [ABBS].

As every method, NMR has well known limitations. Relatively low number of atomic nuclei have enough NMR sensitivity to be easily detectable. Protons have excellent sensitivity so that the list of common liquids or flowing material that can be studied is large. One of the most important limitations of NMR is its incompatibility with ferromagnetic objects as well as the opacity of NMR to electrical conductors. For many years, the time averaging for NMR experiments was larger compared to other techniques, making it more appropriate for studying liquid systems in which the average behavior changed relatively slowly. However, individual spins do not have to attain steady state motion as long as the ensemble behavior can be defined, as shown by studies of turbulence by NMR [dG]. With the development of fast velocity imaging techniques (see for instance [BKZ]) this limitation was removed, reducing the experimental time to the sub-second time scale.

2.2.2 Free Diffusion

Let's consider a liquid where the concentration in number of particles per unit volume is $c(\mathbf{r}, t)$. The flux of particles (per unit area per unit time) is given by Fick's first law of diffusion:

$$\mathbf{J}(\mathbf{r}, t) = -\mathbf{D}\nabla c(\mathbf{r}, t) \qquad (2.1)$$

where $\mathbf{D}$ is a cartesian tensor, so-called *self-diffusion tensor* ($D_{\alpha\beta}$ where α and β take each of the cartesian directions). The equation can be more clearly written as

$$\begin{pmatrix} J(x,t) \\ J(y,t) \\ J(z,t) \end{pmatrix} = -\begin{pmatrix} D_{xx} & D_{xy} & D_{xz} \\ D_{yx} & D_{yy} & D_{yz} \\ D_{zx} & D_{zy} & D_{zz} \end{pmatrix} \begin{pmatrix} \frac{\partial c(x,t)}{\partial x} \\ \frac{\partial c(y,t)}{\partial y} \\ \frac{\partial c(z,t)}{\partial z} \end{pmatrix} \qquad (2.2)$$

Note that the diagonal elements of $\mathbf{D}$ connect concentration gradients and fluxes in the same direction, while the off-diagonal elements couple fluxes and concentration gradient in orthogonal directions.

Because of the conservation of mass, the continuity theorem applies, and thus,

$$\frac{\partial c(\mathbf{r}, t)}{\partial t} = -\nabla \cdot J(\mathbf{r}, t) \qquad (2.3)$$

In other words, the last equation states that $\partial c(\mathbf{r}, t)/\partial t$ is the difference between the influx and the efflux from the point located at $\mathbf{r}$. Combining Fick's first law and the continuity condition we arrive at Fick's second law of diffusion,

$$\frac{\partial c(\mathbf{r}, t)}{\partial t} = \nabla \cdot \mathbf{D}\nabla c(\mathbf{r}, t) \qquad (2.4)$$

For simplicity in what follows we will be concerned only with isotropic diffusion, which can be described by the isotropic diffusion coefficient D (i.e. a scalar). Thus, the two Fick's laws become,

$$\mathbf{J}(\mathbf{r}, t) = -D\nabla c(\mathbf{r}, t)$$

$$\frac{\partial c(\mathbf{r}, t)}{\partial t} = D\nabla^2 c(\mathbf{r}, t) \qquad (2.5)$$

In the case of self-diffusion, there is no net concentration gradient, and instead we are concerned with the total probability $P(\mathbf{r}_1, t)$ of finding a particle at position $\mathbf{r}_1$ at time t. $P(\mathbf{r}_1, t)$ is a sort of ensemble-averaged probability concentration for a single

2.2. Diffusion and Flow in NMR

particle, and it is thus reasonable to assume that it obeys the diffusion equation [Cal]. This is given by

$$P(\mathbf{r}_1, t) = \int \rho(\mathbf{r}_0) P(\mathbf{r}_0, \mathbf{r}_1, t) d\mathbf{r}_0 \qquad (2.6)$$

where $\rho(\mathbf{r}_0)$ is the particle density, and thus, $\rho(\mathbf{r}_0) P(\mathbf{r}_0, \mathbf{r}_1, t)$ is the probability of starting from $\mathbf{r}_0$ and moving to $\mathbf{r}_1$ in time t. The integration over $\mathbf{r}_0$ accounts for all possible starting points. Because the spatial derivatives in Fick's laws refer to $\mathbf{r}_1$ we can rewrite those laws in terms of $P(\mathbf{r}_0, \mathbf{r}_1, t)$ with the initial condition,

$$P(\mathbf{r}_0, \mathbf{r}_1, 0) = \delta(\mathbf{r}_1 - \mathbf{r}_0) \qquad (2.7)$$

where δ denotes the Dirac delta function. Thus, if in Fick's first law $P(\mathbf{r}_0, \mathbf{r}_1, t)$ is substituted for $c(\mathbf{r}, t)$, $\mathbf{J}$ becomes the conditional probability flux. Similarly, in terms of $P(\mathbf{r}_0, \mathbf{r}_1, t)$ the second law becomes,

$$\frac{\partial P(\mathbf{r}_0, \mathbf{r}_1, t)}{\partial t} = D \nabla^2 P(\mathbf{r}_0, \mathbf{r}_1, t). \qquad (2.8)$$

For the case of three dimensional diffusion in an isotropic and homogeneous medium, i.e. boundary condition $P \to 0$ as $\mathbf{r}_1 \to \infty$, $P(\mathbf{r}_0, \mathbf{r}_1, t)$ can be determined by solving eqn. (2.8) with the initial condition given in eqn. (2.7), yielding

$$P(\mathbf{r}_0, \mathbf{r}_1, t) = (4\pi D t)^{-3/2} \exp\left(-\frac{(\mathbf{r}_1 - \mathbf{r}_0)^2}{4 D t}\right) \qquad (2.9)$$

It states that the radial distribution function of the spins in an infinitely large system is Gaussian. Note that $P(\mathbf{r}_0, \mathbf{r}_1, t)$ does not depend on the initial position $\mathbf{r}_0$, but depends only on the net displacement $\mathbf{r}_1 - \mathbf{r}_0$, often referred to as dynamic displacement $\mathbf{R}$. This reflects the Markovian nature of Brownian motion. The solution of eqn. (2.8) becomes much more complicated when the displacement of the particle is affected by its boundary conditions, and $P(\mathbf{r}_0, \mathbf{r}_1, t)$ is not longer Gaussian. The solutions for many cases of interest can be found in the literature [VK, Cra].

The mean-squared displacement is given by

$$\overline{(\mathbf{r}_1 - \mathbf{r}_0)^2} = \int_{-\infty}^{+\infty} (\mathbf{r}_1 - \mathbf{r}_0)^2 P(\mathbf{r}_1, t) d\mathbf{r}_1 = \int_{-\infty}^{+\infty} (\mathbf{r}_1 - \mathbf{r}_0)^2 \rho(\mathbf{r}_0) P(\mathbf{r}_0, \mathbf{r}_1, t) d\mathbf{r}_0 d\mathbf{r}_1 \qquad (2.10)$$

Using eqn. (2.9) we can calculate the mean-squared displacement of free diffusion, giving,

$$\overline{(\mathbf{r}_1 - \mathbf{r}_0)^2} = 2nDt \qquad (2.11)$$

where n is the dimension in which the diffusion in considered (i.e. 1, 2 or 3). Equation (2.10) represents the connection between the molecular displacement due to diffusion and the diffusion equation. Specifically for free diffusion, it states that the mean squared displacement changes linearly with time (i.e. eqn. (2.11)).

For completeness we now make a simple extension to eqn. (2.9) which applies when the diffusion is superposed on flow of velocity $\mathbf{v}$. In this case, a term $\nabla \cdot \mathbf{v} P$ must be added to the right-hand side of eqn. (2.8) to give,

$$\frac{\partial P(\mathbf{r}_0, \mathbf{r}_1, t)}{\partial t} = D\nabla^2 P(\mathbf{r}_0, \mathbf{r}_1, t) + \nabla \cdot \mathbf{v} P(\mathbf{r}_0, \mathbf{r}_1, t). \tag{2.12}$$

If $\mathbf{v}$ is constant then the solution is

$$P(\mathbf{r}_0, \mathbf{r}_1, t) = (4\pi D t)^{-3/2} \exp\left(-\frac{([\mathbf{r}_1 - \mathbf{r}_0] - \mathbf{v}t)^2}{4Dt}\right) \tag{2.13}$$

Note that $P(\mathbf{r}_0, \mathbf{r}_1, t)$ is a normalized Gaussian function of dynamic displacement $\mathbf{R} = \mathbf{r}_1 - \mathbf{r}_0$ with width increasing as time advances.

When being concerned only with motions along one dimension (e.g. z−direction), it is helpful to write the propagator in Cartesian component form and integrating over the remaining two dimensions, to obtain

$$P(Z, t) = (4\pi D t)^{-1/2} \exp\left(-\frac{(Z - v_z t)^2}{4Dt}\right) \tag{2.14}$$

where $Z = z_1 - z_0$ and $v_z = \mathbf{v} \cdot \hat{k}$ (the unitary vectors indicating the three cartesian directions x, y and z, are denoted here by $\hat{i}$, $\hat{j}$ and $\hat{k}$). [Cal]

2.2.3 Velocity Correlation and Self-Diffusion Tensor

The function $P(\mathbf{r}_0, \mathbf{r}_1, \Delta)$ helps us to calculate averages relative to the ensemble of spins. Sometimes a problem naturally lends itself to a description in terms of this function but in other situations the connection is not so obvious. One alternative approach in dealing with behavior which fluctuates with time is to define families of autocorrelation functions. Suppose we have some molecular quantity A which is a function of time. The autocorrelation function of A is [VK]

$$G(t) = \int_0^\infty A(t')A^*(t'+t)dt' \tag{2.15}$$

In a stationary ensemble the average over time implied by eqn. (2.15) could equally be an average over the particles in the ensemble since one particle is representative

2.2. Diffusion and Flow in NMR

of all particles over a sufficiently long time interval. Incorporating these ideas we will rewrite $G(t)$ as

$$G(t) = \overline{A(0)A^*(t)} \qquad (2.16)$$

In effect $G(t)$ measures the rate at which $A(t')$ "loses memory" of its previous values. The time-scale for this loss of memory is called the *correlation time*, τ_c and is defined by

$$\tau_c = \frac{\int_0^\infty \overline{A(0)A^*(t)} dt}{\overline{A(0)^2}} \qquad (2.17)$$

In some experiments it is the spectrum, or Fourier transform, of $G(t)$ which is important. In translational motion theory the spectrum of the velocity correlation function is the self-diffusion tensor [Ste], $D_{\alpha\beta}(\omega)$. Hence,

$$D_{\alpha\beta}(\omega) = \frac{1}{2} \int_{-\infty}^{+\infty} \overline{v_\alpha(0)v_\beta(t)} \exp(i\omega t) dt \qquad (2.18)$$

Generally we shall not be concerned with correlations between differing components of $\mathbf{v}(t)$. We shall therefore focus on the diagonal elements of D and, using the even property of $G(t)$, we write (for the case of z-direction)

$$D_{zz}(\omega) = \int_0^{+\infty} \overline{v_z(0)v_z(t)} \exp(i\omega t) dt \qquad (2.19)$$

This equation tells us that the zero-frequency component of D is simply the time integral of the velocity correlation function, i.e.

$$D_{zz}(0) = \int_0^{+\infty} \overline{v_z(0)v_z(t)} dt \qquad (2.20)$$

By the definition given in eqn. (2.17), results

$$D_{zz}(0) = \overline{v_z^2} \tau_c \qquad (2.21)$$

2.2.4 Pulsed Gradient Spin-Echo NMR

In the presence of a spatially homogeneous $\mathbf{B}_0$ static magnetic field, the spins within the sample will precess with a frequency related to the strength of that magnetic field by the well known Larmor equation:

$$\omega_0 = \gamma B_0 \qquad (2.22)$$

where ω_0 is the Larmor frequency, in *radians* s^{-1}, γ is the gyromagnetic ratio of the spin nuclei, in *radians* $T^{-1}s^{-1}$ and $B_0 = |\mathbf{B}_0|$ in T. We consider here $\mathbf{B}_0$ oriented in the z-direction. However, if in addition to $\mathbf{B}_0$ there is a spatially dependent magnetic field gradient $\mathbf{g}$ (in $T\,m^{-1}$), the frequency of the spins becomes spatially dependent,

$$\omega(\mathbf{r}) = \gamma B_0 + \gamma \mathbf{g} \cdot \mathbf{r} \qquad (2.23)$$

In general, magnetic field gradients may be represented by a tensor which describes the variation of three Cartesian components of the magnetic field along the three independent Cartesian axes. Because such gradients mostly result in additional magnetic fields much smaller than the polarizing field normally used in NMR, it is conventional to regard the Larmor frequency as being affected solely by components of these fields parallel to the polarizing field axis, since orthogonal components have only the effect of slightly tilting the net field direction. In the following discussion we will be concerned only with the variation of the polarizing field magnitude,

$$\mathbf{g} = \nabla \mathbf{B}_0 = \frac{\partial \mathbf{B}_z}{\partial x} \hat{i} + \frac{\partial \mathbf{B}_z}{\partial y} \hat{j} + \frac{\partial \mathbf{B}_z}{\partial z} \hat{k} \qquad (2.24)$$

The important point is that if a homogeneous gradient of known magnitude is imposed onto the sample, the Larmor frequency becomes a spatial label with respect to the direction of the gradient. PGSE NMR uses two narrow gradient pulses of amplitude $\mathbf{g}$, duration δ, and separation Δ placed in the dephasing and rephasing segments of a spin-echo, as shown in Fig. 2.1. These pulses effectively define the starting and finishing point of spin translational motion over a well-defined timescale, Δ, but the unambiguousness of this definition depends on the assumption that the spins move an insignificant distance during the gradient pulse itself, or, in other words, $\delta \ll \Delta$ and $\delta \ll \tau_c$. This narrow gradient pulse approximation is helpful, since it allows one to use the propagator formalism to describe the result of the PGSE experiment. The effect of the first gradient pulse in the sequence is to impart a phase shift $\gamma \delta \mathbf{g} \cdot \mathbf{r}_0$ to a spin located at position $\mathbf{r}_0$ at the instant of the pulse. This phase shift is subsequently inverted by a 180°_x r. f. pulse. Suppose that the molecule containing the spin has moved to $\mathbf{r}_1$ at the time of the second gradient pulse. The net phase shift following this pulse will be $\gamma \delta \mathbf{g} \cdot (\mathbf{r}_1 - \mathbf{r}_0)$. If the spins are stationary then, of course, a perfectly refocused echo will occur. Any motion of the spins will cause phase shifts in their contribution to the echo. The size of this shift is a product of two vectors, the dynamic displacement $(\mathbf{r}_1 - \mathbf{r}_0)$ and a vector $\gamma \delta \mathbf{g}$.

The echo signal $E_\Delta(\mathbf{g})$ is defined as the amplitude of the echo at its center [Cal]. The total signal is a superposition of transverse magnetizations, an ensemble

2.2. Diffusion and Flow in NMR

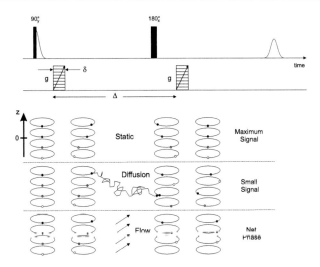

Figure 2.1: PGSE NMR schematically shown, with the gradient pulses of duration δ separated by a much longer time, Δ. At the echo formation, the static spins will have maximum signal, while the spins experiencing diffusion will present a decay due to the random distribution of remnant phases. The spins affected by net flow will contribute with a net phase to the echo, without any decrease in amplitude.

average in which each phase term $\exp\left[i\gamma\delta\mathbf{g}\cdot(\mathbf{r}_1-\mathbf{r}_0)\right]$ is weighted by the probability for a spin to begin at $\mathbf{r}_0$ and move to $\mathbf{r}_1$ in the time interval Δ. This probability is $\rho(\mathbf{r}_0)P(\mathbf{r}_0,\mathbf{r}_1,\Delta)$. Thus

$$E_\Delta(\mathbf{g}) = \int_{-\infty}^{+\infty} \rho(\mathbf{r}_0)P(\mathbf{r}_0,\mathbf{r}_1,\Delta)\exp\left[i\gamma\delta\mathbf{g}\cdot(\mathbf{r}_1-\mathbf{r}_0)\right]d\mathbf{r}_0 d\mathbf{r}_1 \qquad (2.25)$$

So far we did not consider the relaxation processes that occur during the echo sequence. In the absence of diffusion and/or the absence of gradients, the signal at echo time t_E would be given by

$$S(t_E)_{\mathbf{g}=0} = S(0)\exp\left(-\frac{t_E}{T_2}\right) \qquad (2.26)$$

where, for simplicity, we have assumed here that the observed signal results from one species with a single relaxation time. In the presence of diffusion and gradient pulses, the attenuation due to relaxation and the attenuation due to diffusion and the applied gradient pulses are independent of each other, and so we can write,

$$S(t_E) = S(0)\exp\left(-\frac{t_E}{T_2}\right) f(\delta, \mathbf{g}, \Delta, D) \tag{2.27}$$

where $f(\delta, \mathbf{g}, \Delta, D)$ is a function that represents the attenuation due to diffusion, including all the variables involved [Pri]. Thus, if the PGSE measurement is performed keeping t_E constant, it is possible to separate the contributions. Hence, by dividing the signal with gradients by the signal which considers only relaxation attenuation, we normalize out the relaxation effect, leaving only the attenuation due to diffusion

$$E_\Delta = \frac{S(t_E)}{S(t_E)_{\mathbf{g}=0}} = f(\delta, \mathbf{g}, \Delta, D) \tag{2.28}$$

In view of this short discussion, we can keep considering only the attenuation caused by diffusion.

By means of the definition of the reciprocal space vector [Cal] $\mathbf{q}$

$$\mathbf{q} = (2\pi)^{-1}\gamma\mathbf{g}\delta \tag{2.29}$$

eqn. (2.25) can be expressed as

$$E_\Delta(\mathbf{q}) = \int_{-\infty}^{+\infty} \rho(\mathbf{r}_0) P(\mathbf{r}_0, \mathbf{r}_1, \Delta) \exp\left[i2\pi\mathbf{q} \cdot (\mathbf{r}_1 - \mathbf{r}_0)\right] d\mathbf{r}_0 d\mathbf{r}_1 \tag{2.30}$$

The expression $\int \rho(\mathbf{r}_0) P(\mathbf{r}_0, \mathbf{r}_1, \Delta) d\mathbf{r}_0$ defines the average propagator [KH], the probability that a molecule at *any starting position* is displaced by $\mathbf{R} = \mathbf{r}_1 - \mathbf{r}_0$ over time Δ. Consequently, eqn. (2.30) can be written [Cal]

$$E_\Delta(\mathbf{q}) = \int_{-\infty}^{+\infty} \overline{P}(\mathbf{R}, \Delta) \exp\left[i2\pi\mathbf{q} \cdot \mathbf{R}\right] d\mathbf{R} \tag{2.31}$$

Notice that the Fourier transformation of the echo attenuation with respect to $\mathbf{q}$ returns the average propagator for nuclear spin displacement, $\overline{P}(\mathbf{R}, \Delta)$. Considering, once again, the case of performing the experiment in a single direction, say z, the component of the dynamic displacement is given by $Z = \mathbf{q} \cdot \mathbf{R}$.

In its original conception, the PGSE NMR experiment was used to measure molecular self-diffusion for which the average propagator is the Gaussian function

$$\overline{P}(Z, \Delta) = (4\pi D\Delta)^{-1/2} \exp\left(-\frac{Z^2}{4D\Delta}\right) \tag{2.32}$$

Comparing eqn. (2.14) (with $v_z = 0$) and eqn. (2.32) it can be noticed that, in case of isotropic diffusion, the average propagator equals the conditional probability

2.2. Diffusion and Flow in NMR

($\overline{P}(\mathbf{R},t) \equiv P(\mathbf{R},t)$), due to the fact that the conditional probability is independent of the starting positions.

By combining eqns. (2.32) and (2.31) the echo attenuation due to diffusion can be calculated for the PGSE NMR experiment, to give

$$E(q, \Delta) = \exp(-4\pi^2 q^2 D \Delta) \qquad (2.33)$$

where $q = |\mathbf{q}|$. This solution holds for the narrow gradient pulse approximation, while for finite gradient pulse durations there is a $\delta/3$ correction to the displacement time, giving

$$E(q, \Delta) = \exp(-4\pi^2 q^2 D (\Delta - \delta/3)) \qquad (2.34)$$

the well-known Stejskal-Tanner relation. The self-diffusion coefficient is then derived from the slope of semilogarithmic plots of E versus $\gamma^2 g^2 \delta^2 (\Delta - \delta/3)$.

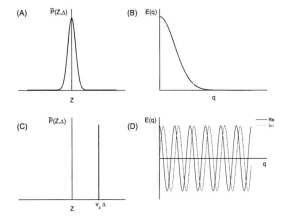

Figure 2.2: The echo attenuation $E(q, \Delta)$ resulting from the PGSE experiment, is the Fourier transform of the probability that a particle will be displaced a distance Z in a time Δ. The diffusive Gaussian propagator (A) will produce an echo attenuation that is also Gaussian (B), while a fluid moving at a constant velocity (C) will affect the phase of the echo (D) but not its magnitude.

In the case of uniform flow, with velocity component v along the direction q, the expressions corresponding to eqns. (2.32) and (2.33) are,

$$\overline{P}(Z,\Delta) = \delta(Z - v\Delta) \tag{2.35}$$

and

$$E(q,\Delta) = \exp(-i2\pi qv\Delta) \tag{2.36}$$

where δ in eqn. (2.35) is the Dirac delta function (not to be confused with the length of the gradient pulses). Note that mean motion modulates the phase of the echo function, while random diffusive motion attenuates the signal amplitude.

In diffusion NMR and flow NMR experiments it is common to analyze the PGSE echo data directly in the domain in which it is collected, the **q**-space. In some cases, Fourier transforming the **q**-space data, which yields a propagator for motion, may provide an intuitive way of representing the physics of a system. However, there are many PGSE NMR experiments where there are advantages to measuring parameters directly from the **q**-space data.

The process of Fourier transforming data induces ringing artifacts if the data are truncated before the signal has dropped below the noise level of the experiment. For this reason the data must be collected over a wide range of the **q**-space. For many cases, such detailed sampling of **q**-space is unnecessary, as the important features of a propagator can be determined readily from a few low **q**-space data points without subsequent Fourier transformation, thus reducing the necessary gradient strength and the experimental time [CCS].

The echo attenuation function described in eqn. (2.31) represents the ensemble-averaged phase shift, $\overline{\exp[i2\pi \mathbf{q} \cdot \mathbf{R}]}$. This expression can be expanded into a Taylor series,

$$E_\Delta(\mathbf{q}) \approx 1 - (1/2!)(2\pi q)^2 \overline{Z^2} + (1/4!)(2\pi q)^4 \overline{Z^4}... \tag{2.37}$$

where, as previously defined, Z is the component of displacement along the gradient direction defined by **q**. Eqn. (2.37) is useful, since it tells us that whatever the shape of the propagator function, $\overline{P}(Z,\Delta)$, the initial decay of $E(q,\Delta)$ with respect to q will always yield the ensemble-averaged mean square displacement $\overline{Z^2}$, and hence an **effective diffusion (or dispersion) coefficient**, $D_{eff} = \overline{Z^2}/2\Delta$

2.2.5 Stationary and Time Dependent Random Flow

Self diffusion in simple liquids provides an example of translational motion which is randomly directed and rapidly fluctuating, with correlation lengths on the order of

10^{-10} m and correlation times very much shorter than the Larmor period. None the less, small molecule stochastic motions can exist at greater length- and time-scales, fluid turbulence being an obvious example.

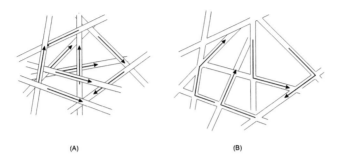

(A)   (B)

Figure 2.3: (A) Example of stationary random flow. The particles move in randomly distributes directions, and the motion is describable by a time-independent velocity field. (B) Example of pseudodiffusion. Flow occurs in branched capillaries so that the velocity direction fluctuates with a correlation time of the order the ratio of the mean branch separation to the mean velocity.

On the microstructural scale irregular motion is a feature of flow through porous media and biological tissue. Here the molecules follow paths which are randomly directed because of the complexity of capillary organization. Fluctuations in this motion may occur because of the random-walk character of the channels and capillaries and because of path divergency at branch points where a small change in initial condition can lead to widely differing outcomes, a classic feature of chaotic behavior. This type of incoherent motion is sometimes called *perfusion*.

In characterizing random flow it is helpful to define two regimes. In the first, which we shall call *stationary random flow* the molecular motions are randomly directed in magnitude and/or orientation and described by a velocity field $\eta(\mathbf{u}, \mathbf{r})$ which is time independent. This is the motion which might be associated with laminar flow in shear or with flow in an array of randomly directed capillaries in which the local director is fixed. Such motion is illustrated in Fig. 2.3A. In the second, which we label *pseudodiffusion*, the molecular velocities are not only randomly distributed across the ensemble but fluctuate in time as well. Examples of this case include turbulence and branched capillary motion as shown in Fig. 2.3B [Cal].

Stationary Random Flow

Here we will consider a liquid or fluid consisting in regions labelled by the index i. The vector $\mathbf{v}$ will be used to describe the mean velocity of the molecules and $\mathbf{u}_i$ to describe the local variation about the mean for the region i. This motion has correlation time τ_{cu} and correlation length $L_u \sim (\overline{\mathbf{u}^2})^{1/2} \tau_{cu}$. Within the region i the individual molecules, labelled by j, will experience stochastic motion $\mathbf{r}_j$ due to self-diffusion. This motion has correlation time τ_{cr} with correlation length $L_r \sim (D\tau_{cr})^{1/2}$. The three motions will be so separated in correlation length and time that we can treat them as stochastically independent. Figure 2.4 shows this model schematically.

The total displacement of a molecule labelled by j in fluid element i is therefore

$$\mathbf{r}_{ij}(t) = \mathbf{v}t + \mathbf{u}_i t + \mathbf{r}_j(t) \tag{2.38}$$

The j-ensemble average over a time t long compared with the molecular diffusive

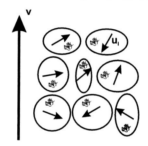

Figure 2.4: General model for a moving fluid. It is supposed to be consisting of regions labelled by the index i, with an associated internal constant velocity $\mathbf{u}_i$. The regions are superimposed onto a mean velocity $\mathbf{v}$ common to all of them. Within the region i the individual molecules, labelled by j, will experience stochastic motion $\mathbf{r}_j$ due to self-diffusion.

correlation time but short compared with the fluid element correlation time, i.e. $\tau_{cr} \ll t \ll \tau_{cu}$ gives $\overline{\mathbf{r}_i} = \mathbf{v}t + \mathbf{u}_i t$ while the average over i for $\tau_{cu} \ll t$ gives $\overline{\mathbf{r}} = \mathbf{v}t$. We shall evaluate the PGSE experiment in the stationary random flow regime, $\tau_{cr} \ll t \ll \tau_{cu}$, where $\mathbf{u}_i$ is time independent.

The PGSE phase shift for a spin being located in molecule ij is,

$$\phi_{ij}(t) = \gamma \int_0^t t' \mathbf{g}^*(t') \cdot (\mathbf{v} + \mathbf{u}_i) dt' + \gamma \int_0^t \mathbf{g}^*(t') \cdot \mathbf{r}_j(t') dt' \tag{2.39}$$

2.2. Diffusion and Flow in NMR

where $\mathbf{g}^*$ represents the effective gradient, which takes into account the r.f. pulses used in the sequence. Every 180° pulse inverts the sign of the effective gradient, while 90° pulses transform it to zero. Fig. 2.5 shows the effective gradient in a PGSE sequence to illustrate its use.

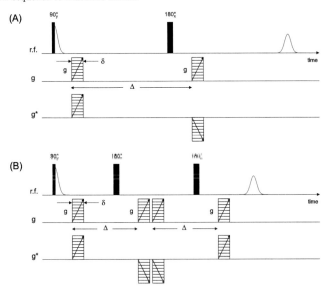

Figure 2.5: PGSE (A) and Double PGSE (B) sequences with different r.f. pulses, gradient pulses, and the effective gradient of the sequence, referred as $\mathbf{g}^*$.

The separability of the i and j averages means that diffusion and random flow are uncorrelated and the averages over i and j are separable. Thus,

$$E(t) = \overline{\exp[i\phi_j(t)]} = \overline{\exp\left[i\gamma \int_0^t \mathbf{g}^*(t') \cdot \mathbf{r}_j(t')dt'\right]} \ \overline{\exp\left[i\gamma \int_0^t t'\mathbf{g}^*(t') \cdot (\mathbf{v} + \mathbf{u}_i)dt'\right]} \quad (2.40)$$

The first factor is the usual diffusive attenuation, equivalent to the exponential term shown in eqn. (2.34). The second differs slightly from that which is usually associated with flow since eqn. (2.40) deals with fluctuating flow. This latter factor is separable into a phase shift due to net flow and a term due to stationary random flow. Defining $\mathbf{p} = \gamma \int_0^t t'\mathbf{g}^*(t')$, usually called 1st moment of gradient, we may write this factor $\exp(i\mathbf{p}\cdot\mathbf{v}) \ \overline{\exp(i\mathbf{p}\cdot\mathbf{u}_i)}$. In the case of the Stejskal-Tanner PGSE sequence (see Fig. 2.5A) the echo attenuation becomes,

$$E(\mathbf{g}) = \exp(i\gamma\delta\mathbf{g}\cdot\mathbf{v}\Delta)\,\overline{\exp(i\mathbf{p}\cdot\mathbf{u}_i)}\,\exp[-\gamma^2\delta^2g^2D(\Delta-\delta/3)] \qquad (2.41)$$

where D is the molecular self-diffusion coefficient. The stationary random flow factor can be evaluated by using the idea that the motion behaves like diffusion but with an observational time-scale very much shorter than the diffusive correlation time [Cal]. In the case of PGSE the exponent becomes $(1/6)\overline{u^2}\gamma^2g^2\delta^2\Delta^2$ [NC], implying that

$$D_{eff} = \frac{1}{6}\overline{u^2}\Delta \qquad (2.42)$$

Stationary random flow is identified by a specific signature in a PGSE experiment: the echo is attenuated in a manner similar to diffusion but with an effective diffusion coefficient proportional to the observation time.

On the other hand, for the Double PGSE (Fig. 2.5B), which is usually referred to as *velocity compensated sequence* and obeys $\mathbf{p}=0$, there is no extra attenuation where the motion remains coherent during the echo formation period.

Pseudodiffusion

By contrast with the stationary random flow regime, pseudodiffusion results when the observational time-scale is sufficiently long that $\tau_{cu} \ll t$. Given eqn. (2.21) we may write down the effective diffusion coefficient as measured by the PGSE experiment,

$$D_{eff} = \overline{u_z^2}\tau_{cu} = \frac{1}{3}\overline{u^2}\tau_{cu} \qquad (2.43)$$

In such an experiment pseudodiffusion behavior is characterized by echo attenuation more severe than that expected from self-diffusion alone. On the other hand, pseudodiffusion attenuation will not refocus in case of using Double PGSE (where $\mathbf{p}=0$).

The Intermediate Case

In case that $t \sim \tau_{cu}$, the signature of the echo attenuation is partially refocusing in the Double PGSE sequence, while the PGSE single echo is strongly attenuated [Cal].

2.3 Reaction Monitoring

2.3.1 Brief Description

In this section, results concerning the reaction evolution are presented, obtained by means of the time dependence of average propagator's shape and effective diffusion coefficient. All the experiments presented here were performed with a single catalyst pellet that was immersed in a 7 mm inner diameter tube, filled with 1.1 ml of 5 % v/v H_2O_2 solution, placed in a Bruker DSX200 spectrometer equipped with a 10 mm birdcage r.f. coil. The device operates at Larmor frequency of 200 MHz. The gradient used was aligned to the axis of gravity in one case, parallel to the external magnetic field and labelled as z−direction here, and perpendicular to that axis in other case, labelled as x−direction.

Two different catalyst particles were used,

(i) a cylindrical pellet with 4 mm diameter and 4 mm height, the matrix being made of Al_2O_3, with Cu as catalytically active sites (Bayer CH-FCH-RD Geb. 8, Cu cont. 51.2 % weight).

(ii) a cylindrical pellet with 3 mm diameter and 3 mm height, also made of Al_2O_3, with Pt as catalytically active sites (Company Alfa Aesar).

The aqueous hydrogen peroxide solutions were obtained by mixing distilled water (CHROMASOLV Plus, Sigma-Aldrich), with hydrogen peroxide 30 % by weight (Riedel-deHaen). The different reaction experiments were monitored without any further hydrogen peroxide supply for several hours. The tube was placed into the coil in a position which defines the sensitive volume close to the pellet, as shown in Fig. 2.6. Then, the signal results from the liquid at maximum 3 mm from the catalyst outer surface. In order to allow comparison between different realizations, the amount of liquid and initial concentrations were kept constant. In both propagators and diffusion experiments, a variant of the PGSE and Double PGSE was used: the Spin Echo was replaced by a Stimulated Echo. This is a rather standard technique which exploits the fact that in many samples T_2 is considerably shorter than T_1. This is the case of H_2O_2 where at 200 MHz Larmor frequency, $T_1 \sim 3$ s while T_2 covers a wide temporal range from 10 ms to 3 s (see next chapter). Figure 2.7 shows the sequences for the case of PGSTE (A) and Double PGSTE (B). The main difference respect to the spin-echo sequences, is the splitting of the 180° pulse into two 90° pulses. Thus, in between the second and third pulses the magnetization

24   Chapter 2. Mass Transport during H_2O_2 Decomposition

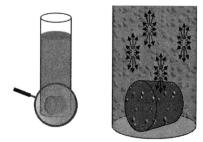

Figure 2.6: Layout used in the experiments. In the right side a magnification of the vicinity of the pellet is shown.

is stored along the z-axis. The third pulse recalls the magnetization to the $x - y$ plane, and a stimulated echo is formed after a time equal to the separation between the first two pulses. The main disadvantage is related with the signal-to-noise ratio, due to the fact that the stimulated echo has half the height than a spin echo.

It should be noted that the stimulated echo pulse sequence also generates two additional spin echoes. These are, respectively, the echo of the initial pulse FID cause by the second pulse, and the echo of the second pulse FID caused by the third pulse. One effective way to avoid them is the use of a homogeneity-spoiling (homospoil) magnetic field gradient applied during the "z-storage" period. This has the effect of destroying the unwanted transverse magnetization without influencing the magnetization which has been stored along the z-axis [Cal].

As mentioned in section 2.1, the metal within the pore space catalyzes the reaction. The rate of oxygen generation is then much higher than the mass transport due to pure molecular diffusion, and the maximum solubility of oxygen in the liquid is reached quickly. That leads to the generation of bubbles. The production and motion of gas bubbles results in a random change in the velocities of the liquid around the pellet, which is displaced and driven by the rising bubbles. The faster the production of gas, the larger the amount of rising bubbles will be; in consequence, the increase of the effective diffusion coefficient (D_{eff}) will be more pronounced. This effect enhances the mass transport inside and outside the pellet, then the rate of reaction. The reaction will accelerate. Moreover, due to the predominantly rising bubbles, a much more increased mean-square displacement is

2.3. Reaction Monitoring

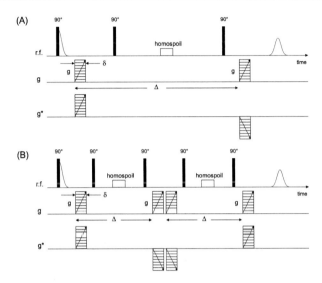

Figure 2.7: Pulse Gradient Stimulated Echoes used in the experiments. (A) PGSTE and (B) Double PGSTE. The main difference respect with to the spin-echoes sequences is the splitting of the 180° pulse into two 90° pulses, and the inclusion of homospoil gradients to avoid the extra spin echoes generated.

expected in $z-$direction compared to $x-$direction. Figure 2.6 shows schematically the situation.

On the other hand, the rate of oxygen generation and, eventually, of bubble production, depends on the H_2O_2 concentration. Based on that, a decrease in the effective diffusion coefficient while the reaction proceeds is expected.

2.3.2 Average Propagators During the Reaction

The 1-D average propagator represents the probability that a molecule starting at any position within the sample, is displaced by a quantity Z (in case of $z-$direction) over the observation time Δ (see eqn. (2.32)). In that interval, the mean velocity $v_z = Z/\Delta$ can be defined, and so the average propagator becomes a velocity probability. Valuable information about the system can be extracted, in many cases, from the 1-D propagator's shape in different directions. In Fig. 2.8, results from

26   Chapter 2. Mass Transport during H$_2$O$_2$ Decomposition

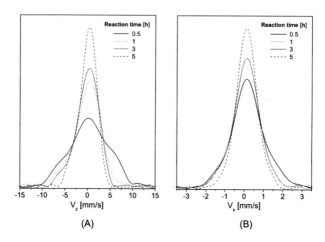

Figure 2.8: Average Propagators corresponding to 5 % v/v of H$_2$O$_2$ reacting in the presence of a Cu-pellet. They were acquired with a Pulse Gradient Stimulated Echo sequence, with the gradient system aligned to (A) z−direction (B) x−direction. In this case, z−positive means up.

experiments performed in z− and x−direction as a function of reaction time are presented. The sample used in this case was the Cu-pellet, and the propagators were obtained by Fourier transforming the echo decay resulting from PGSTE sequence with respect to the wave vector q, similar to that shown in Fig. 2.7A, but the gradient pulses are stepped here from $-g_{max}$ to $+g_{max}$. In both sets of experiments, i.e. both directions, the observation time was set to be $\Delta = 45$ ms, with $\delta = 1$ ms and 32 gradient steps, with 2 scans per step. Every single propagator was acquired in 4.5 minutes.

Each plot shows propagators for 4 different times: 1/2, 1, 3 and 5 hours respectively. Close to the beginning, the propagator corresponding to the z−direction is asymmetric, with higher positive velocities being slightly more probable, where positive here means up. However, after 1 hour of reaction it becomes symmetric, showing a Gaussian-like shape. On the other hand, in x−direction, even at the beginning the propagators are almost symmetric, the highest velocities (~ 3 mm/s) being much smaller than in the previous case (~ 10 mm/s). In both directions,

2.3. Reaction Monitoring

as expected, all the propagators are centered in zero, because no mean velocity is present.

The propagator, as any other probability function, has unitary integral. So, as the reaction advances and both H_2O_2 concentration and rate of bubbles generation decrease, the propagators become narrower and higher. After 5 hours, it can be seen that the maximum velocity in $z-$direction is still about twice that of corresponding to $x-$direction.

Notice that in $x-$direction, the propagators corresponding to 1/2 and 1 hours do not present much change, which renders the latter difficult to observe in Fig. 2.8B.

Although this is a relatively standard technique in studying fluid dynamics, in this case, as pointed out in the last section, it presents some disadvantages, because the information about the reaction evolution is not directly read from the collection of propagators. Despite the fact that a major number of points must be acquired in **q**$-$space, from negative to positive, and increased gradient strengths must be used to make the echo decay to reach the noise level of the experiment, in order to avoid ringing effects following the Fourier transformation. In cases like this, it is preferable to monitor the effective diffusion coefficient during the reaction. Such measurements only requires to acquire the first few points of the decay, reducing maximum gradient strength as well as experimental time (see the discussion accompanying eqn. (2.37)).

2.3.3 D_{eff}^z and D_{eff}^x vs. Reaction Time During the Decomposition with Cu-Pellet

The same reaction conditions were employed to follow, via monitoring the effective diffusion coefficient, the H_2O_2 decomposition in the presence of, once more, the catalyst particle doped with Cu. The experiments were performed in both $z-$ and $x-$direction alternated during the same decomposition, i.e. the whole experiments consisted of a series of $D_{eff}^z - D_{eff}^x - D_{eff}^z - D_{eff}^x - ...$ measurements, with the pulse sequence shown in Fig. 2.7A. The observation time was set to be $\Delta = 45\ ms$, equal for both directions, with $\delta = 1\ ms$ and 20 gradient steps, with 4 scans per step. Every single experiment was acquired in 2.5 minutes. Figure 2.9 shows the results for a reaction time of about 17 hours. The quotient D_{eff}/D_0 for both directions is plotted, where D_0 represents the isotropic diffusion coefficient measured in water in the presence of the catalyst particle, the condition expected at the end of the reaction.

28   Chapter 2. Mass Transport during H$_2$O$_2$ Decomposition

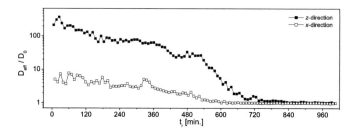

Figure 2.9: Evolution of the effective diffusion coefficient (normalized by the molecular diff. coeff., D_{eff}/D_0) in $z-$ and $x-$direction vs. reaction time, in an H$_2$O$_2$ decomposition catalyzed by a pellet doped with Cu, during ~ 17 hours. The PGSTE sequence was used, with observation time $\Delta = 45$ ms

The plot contains similar information to the one presenting propagators in Fig. 2.8, but here the experimental time was much shorter, even when the data were acquired with twice the number of scans. The information is much easier to read as well. The reaction starts with values more than two orders of magnitude larger in $z-$ and about one order of magnitude larger in $x-$direction, compared to solely isotropic self-diffusion. During approximately the first 9 hours the decay presents some regular behavior. In that period, the reaction rate at the catalyst sites is high enough to produce bubbles continuously, in a wide variety of sizes, which perturb the liquid and enhance the transport of H$_2$O$_2$ molecules to the pellet. Notice that the values in both directions follow the same trend. At $t_r = 530$ minutes (~ 8.8 hours), the $x-$component is close to the self-diffusion value which means that from this point on, the rate of reaction is too low compared to the initial values, leading to a much lower rate of bubbles generation. Hence, the mass transport dramatically decreases, which is translated in a much more pronounced decrease of the bubble's generation rate. These two values are coupled, affecting each other. As a result the effective diffusion coefficient starts a period of more pronounced decay ($z-$direction). About $t_r = 14$ hours (840 minutes), both components coalesce in the unique isotropic self-diffusion value. Nevertheless, it is not possible to conclude that the decomposition has finished. It might happen that beyond this time the rate of decomposition is

2.3. Reaction Monitoring

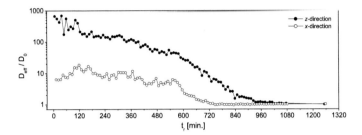

Figure 2.10: Evolution of D_{eff}/D_0 in $z-$ and $x-$direction vs. reaction time, measured under the same conditions as curves presented in Fig. 2.9, except for the observation time, which was set to be $\Delta = 90$ ms

very low, permitting the molecular diffusion to transport the H_2O_2 to the catalyst and the extra oxygen away from the pellet, without forming bubbles. Another possibility is that due to the volume of liquid, little bubbles generated by extremely low reaction rates slightly contribute in increasing the bulk self-diffusion coefficient, being effectively imperceptible.

The latter arguments are reinforced by Fig. 2.10 where the same experiment is plotted, measured setting $\Delta = 90$ ms. It is expected that using longer observation times the sequence becomes more "sensitive" to motion (see eqn. (2.42)), translating it in a much increased echo decay. It can be clearly observed that, at any t_r value, the effective diffusion coefficient is bigger compared to the case of $\Delta = 45$ ms. So, even at reaction times where the bubbles slightly perturb the liquid around the pellet, making the effect impossible to observe in Fig. 2.9, it is perfectly observable in Fig. 2.10. Despite the first two hours when the decay shows some oscillations in $z-$direction and it increases in $x-$direction, there is a regular decay until $t_r = 580$ minutes while in the previous case it corresponded to 530 minutes. Notice that, at that time, D^x_{eff} is still about one order of magnitude larger than the limiting value, in contrast with the case of Fig. 2.9.

The lower value of D^x_{eff} is reached at $t_r = 750$ minutes, while for D^z_{eff} this time corresponds to ~ 1000 minutes. It is also noticed that between $t_r = 580$ and 750 minutes, D^z_{eff} and D^x_{eff} follow the same trend.

In other words, during a long period of time, the rate of mass transport to the catalyst enhanced by the rising bubbles is much larger than the rate of H_2O_2 reaction at the sites, and concentration decreases together with the effective diffusion coefficient following a regular trend. However, at a certain concentration value, or equivalently at a certain rate of bubble production, the rate of reaction at the metal sites and the mass transport rate become comparable. The decomposition starts to be diffusion controlled. Then less and smaller bubbles are produced and the effective diffusion coefficient decreases. This effect is observed at different reaction times depending on the Δ value used. The longer the observation time, the smaller perturbation the sequence is able to reflect into the echo decay.

2.3.4 D_{eff}^z vs. Δ during the Decomposition with Cu-pellet

In 2.2.5 the cases of stationary random flow and pseudodiffusion were treated in detail. It has been pointed out that, in the former case, the corresponding effective diffusion coefficient obtained by the first part of the echo decay is dependent on Δ, in contrast with the pseudodiffusion case. Experiments with different observation times were then performed, in order to estimate the correlation time of the liquid shaken by the bubbles, relative to Δ.

Figure 2.11 shows the results of effective diffusion coefficients measurements in z−direction during the decomposition with a Cu-pellet for a period of about 17 hours (different total t_r for different experiments) with different observation times, $\Delta = 90, 45$, and $10\ ms$. In the last two cases, the width of the gradient pulses was the same, $\delta = 1\ ms$, while in the first case that time was shorter, $\delta = 500\ \mu s$, due to the fact that even the smallest gradient strength reliably produced in the spectrometer led to a very pronounced decay. The same set of data in x−direction (data not shown) presents a similar aspect, with correspondingly much smaller values.

The D_{eff}^z values for the first period of the decay are plotted versus Δ in Fig. 2.12. The lines in the plot represent linear fittings. As can be observed, all sets of values present a linear dependence with Δ. Eqn. (2.42) states that, in case of stationary random flow, D_{eff} should be linearly related with Δ, where the slope is a fraction of the mean- squared random velocity, $\overline{u^2}$. As expected, while the decomposition evolves, the mean velocity of the liquid driven by the bubbles decreases, in accordance with the slopes in Fig. 2.12. However, in case of being in the stationary

2.3. Reaction Monitoring

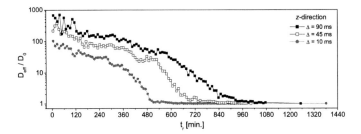

Figure 2.11: D_{eff}/D_0 in $z-$ direction vs. reaction time, obtained with a PGSTE sequence in the same conditions presented in previous plots, for three different observation times, $\Delta = 90, 45$ and $10\ ms$

random flow regime ($\Delta \ll \tau_{cu}$), a Double PGSTE sequence (Fig. 2.7B) must refocus the random contribution from the bubbles (the term related with $\mathbf{u}$), thus giving an effective diffusion coefficient independent of Δ. Moreover, by inspection of eqn. (2.41) it can be noticed that the only remaining term must be that corresponding to isotropic self diffusion, due the fact that we are dealing with a system with $\mathbf{v} = 0$ (i.e. no net flow is present in the tube).

A set of experiments was performed in order to monitor the effective diffusion coefficient in the same sample and conditions presented above, for different observation times, but using the Double PGSTE sequence presented in Fig. 2.7B instead. In this set, the experiments were also carried out in both $z-$ and $x-$direction alternated during the same decomposition, for each Δ. In order to allow comparisons with the previous measurements, as the sequence has two periods of evolution of duration Δ, those values were chosen such that $2\Delta_{\text{Double}} = \Delta_{\text{PGSTE}}$. For the sake of clarity, however, in what follows we will refer to the observation time.

The width of the pulse gradients was set to be $\delta = 500\ \mu s$ and kept constant during all the experiments. The phase cycling for this sequence requires 8 scans [JM] so the number of gradient steps used was 10, in order to keep the acquisition time for individual experiments equal to those performed with a PGSTE sequence, 2.5 minutes. Thus, the temporal resolution during the reaction is the same in both sets.

Figure 2.13A shows the results (in $z-$direction only) obtained during circa 18

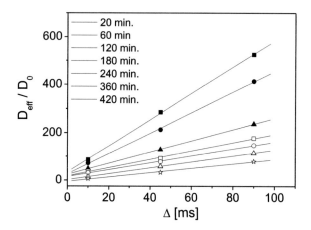

Figure 2.12: D_{eff} vs. Δ in z-direction, corresponding to different stages of the reaction curves presented in Fig. 2.11. The lines represent linear fittings. The decrease in the slopes is related with the drop on the mean-squared velocities $\overline{(u^2)}$ produced by the rising bubbles in the liquid.

hours of reaction for different observation times: 90, 45, and 10 ms. The values differ from that expected by solely molecular diffusion in the case of stationary random flow. Although much lower compared to those obtained in the PGSTE experiments, they allow to follow the reaction progress as well. Moreover, the dependence on the observation time is evident. Figure 2.13B shows D^z_{eff}/D_0 obtained with 45 ms observation time by using either PGSTE or Double PGSTE. By that plot, it is possible to conclude that, in the first 500 minutes of reaction, the liquid motion generated by the bubbles close to the catalyst particle is in the intermediate regime, between stationary random flow and pseudodiffusion, where the observation time and the correlation time are of the same order, $\Delta \sim \tau_{cu}$. After that, The Double PGSTE sequence is effective in refocussing the extra term in eqn. (2.41) related to **u** and then, the motion is in the stationary random flow regime. It means, the smaller amount of bubbles impart a much more coherent motion to the liquid surrounding,

2.3. Reaction Monitoring

resulting in a much longer correlation time τ_{cu}; hence, $\tau_{cu} \gg \Delta$. Given the same

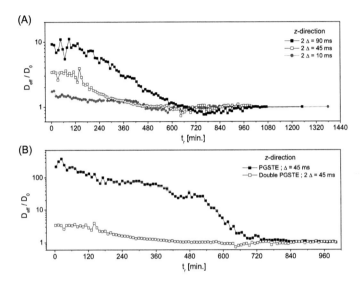

Figure 2.13: (A) D_{eff}/D_0 in z–direction vs. reaction time obtained from H_2O_2 decomposing in the presence of a Cu-catalyst, by means of a Double PGSTE sequence. Three different observation times were used, $90, 45$ and $10\ ms$. (B) Comparison between equivalent experimental data from PGSTE and Double PGSTE with $45\ ms$ observation time.

reaction, like here, as Δ increases τ_{cu} needs more reaction time to greatly exceed that value. It can be seen from the plot in Fig. 2.13A, where the curves reach the asymptote, i.e. self diffusion, for longer reaction times when Δ is longer.

2.3.5 Comparison Between Two Different Catalyst Samples

In order to make use of the effective diffusion coefficient evolution curves to compare different catalysts, a further set of experiments was carried out. The same total amount of liquid, as well as initial H_2O_2 concentration and pulse sequence parameters were used in a decomposition catalyzed by a cylindrical pellet doped with Pt. This particle has proven to be much more active in catalyzing the reaction. Figure 2.14 shows D_{eff}/D_0 in z-direction as a function of reaction time, for both catalyst metals, obtained by means of PGSTE experiments with $\Delta = 45\ ms$. The initial values are much bigger in case of the reaction with the Pt pellet. The effective diffusion coefficients rapidly decrease while the reaction takes place. After

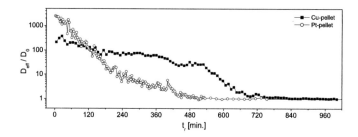

Figure 2.14: D_{eff}/D_0 vs. reaction time obtained by using a PGSTE sequence with $\Delta = 45\ ms$ for the decomposition of H_2O_2 catalyzed by two different catalyst particles, either doped with Pt or Cu.

approximately 300 minutes the reaction has started, there is a change in the slope, associated with a drop in the rate of bubble generation and decomposition. In this case, due to the higher rate of decomposition at the sites compared to the Cu-pellet, it can be assumed that diffusion itself is never sufficient to dissolve the extra oxygen generated in the pore space and bubbles are always present. The reaction never becomes diffusion controlled. Hence, the pronounced decrease of D_{eff} as in the previous case, is not observed here.

Finally, the effective diffusion reaches the final expected value at circa $t_r = 530$ minutes in case of using Pt-pellet, in contrast with $t_r \sim 840$ minutes in case of Cu-pellet.

The relative shape as well as the reaction time values at which the curves presented above show different details, give qualitative and quantitative information about the reaction itself, allowing comparisons between catalyst particles with dif-

2.3. Reaction Monitoring

ferent: porosity, metal content, type of metal, etc. On the other hand, comparison between effective diffusion coefficients obtained with PGSTE and Double PGSTE sequence for different observation times (Δ) gives an estimation of the correlation time of the liquid motion surrounding the catalyst at any time during the reaction. By means of these experiments it is possible, in principle, to correlate different parameters of the catalyst with the size of bubbles and/or their rate of generation. Further details about the extraction of dynamics information from that curves is beyond the scope of this work, which was focused mainly on demonstrating the feasibility of using NMR in such kind of problems. It is necessary to mention, before ending, that this sort of experiments can be straightforwardly extended to the case of reactors including a net flow of the reactants. The extra decay produced by a coherent velocity distribution can be easily subtracted from the reaction effects by performing a first experiment without any reaction, either by replacing the liquid with another one possessing similar viscosity or by replacing the catalyst particle with a free-of-metal pellet with similar dimensions. Other ideas might be implemented, like filtering the much more coherent motion from the flow by means of a Double PGSTE with properly chosen Δ.

Chapter 3

Chemical Exchange and Relaxation in H_2O_2

If you have an apple and I have an apple and we exchange these apples then you and I will still each have one apple. But if you have an idea and I have an idea and we exchange these ideas, then each of us will have two ideas.

George Bernard Shaw.

3.1 Chemical Exchange in NMR

3.1.1 General

The NMR properties of atoms and molecules strongly depend on short-range effects, being then affected by molecular motion. The way in which NMR parameters are affected depends on the time scale of the motion, and how it modifies the nuclear spin hamiltonian. On the other hand, the exchange of ions or molecules between various sites is a common occurrence in chemical and biological systems. Chemical exchange, where the nuclear spins change their environment through a chemical process (which can be intra- or intermolecular) is in the range of nano-seconds to seconds. Such wide range includes many different effects, from changes in the time correlation functions to influences on the dynamics of nuclear magnetization [Wen, Lev2]

Spectral densities are a central concept in NMR relaxation theory. They provide the tool through which measurements of relaxation parameters can be seen as information on molecular dynamics. The relaxation times are expressed in terms of the spectral densities $J_h(\omega)$ of the time correlation function $\overline{F_h(t)F_h^*(t+\tau)}$ of nuclear spin interactions $F_h(t)$ at the angular frequency ω as given by

$$J_h(\omega) = \int_{-\infty}^{\infty} \overline{F_h(t)F_h^*(t+\tau)}\exp(i\omega\tau)d\tau \qquad (3.1)$$

The interaction function $F_h(t)$ is the product of a nuclear relaxation interaction strength (proportional to r^{-3} for magnetic dipolar relaxation, for example) and a function of orientation (second- order spherical harmonics), and contains the information about the molecular environment [Woe3]. Due to the relationship between the interaction function and *rank-2 irreducible spherical tensor operators* as well as second-order spherical harmonics, the subindex h can take only the values 0, 1 or 2.

The correlation function is often assumed to have a single exponential form

$$\overline{F_h(t)F_h^*(t+\tau)} = \overline{F_h(t)F_h^*(t)} \exp(-|\tau|/\tau_c). \qquad (3.2)$$

For instance, the correlation function for the model describing the motion of a rigid spherical molecule in an ordinary, isotropic liquid (i.e. whose macroscopic properties do not depend on orientation) can be expressed as shown above, with the correlation time τ_c depending on the rate of rotational molecular diffusion, $\tau_c = (6D_R)^{-1}$ [KM], where D_R represents the rotational diffusion coefficient. When the time correlation function has an exponential form, the spectral densities $J_h(\omega)$ are given by the Lorentzian functions

$$J_h(\omega) = \overline{F_h(t)F_h^*(t)} \frac{2\tau_c}{1+\omega^2\tau_c^2} \qquad (3.3)$$

Depending on the model used for the interactions, the relaxation times of the system can be obtained. Consider for example, the case of magnetic dipolar interactions between two identical spin-1/2 nuclei. The relaxation times are given by [Abr]

$$\frac{1}{T_1} = \frac{9}{8}\gamma^4\hbar^2 r^{-6} \left[J_1(\omega_0) + J_2(2\omega_0)\right]$$

$$\frac{1}{T_2} = \frac{9}{32}\gamma^4\hbar^2 r^{-6} \left[J_0(0) + 10J_1(\omega_0) + J_2(2\omega_0)\right],$$

where T_1 is the spin-lattice relaxation time, T_2 the transverse relaxation time and ω_0 the Larmor frequency.

In a more general case that includes chemical exchange between different types of sites (designated by k) it is necessary to look closely at the evaluation of the correlation function $\overline{F_h(t)F_h^*(t+\tau)}$ in eqn. (3.2) averaged over the ensemble, in

3.1. Chemical Exchange in NMR

order to calculate NMR relaxation times properly [Wen, Woe3] . For a given nucleus at time t in site k the interaction function $F_h(t)$ must be evaluated. As a second step, the correlation function at times $t + \tau$ needs to be calculated, increasing τ from zero to *large* values, using the correlation time of the site k, τ_{ck}. Finally, taking the sum over all the different type of sites k, weighting each site by its fractional population p_k, the time correlation function is obtained. The averaged time that the nucleus occupies a site of type k is denoted by τ_k, the so-called life times [Wen].

When the chemical exchange is such that all the lifetimes τ_k are long compared with the correlation times of the sites, τ_{ck}, the relaxation times of the nuclei at different sites are independent of exchange. In that case, each site has an associated relaxation time, and the above-mentioned treatment for calculating the spectral densities holds.

In cases when the exchange is extremely fast, so that the lifetimes of the nuclei at the sites are on the order of rotational correlation times (typically nanoseconds to picoseconds), the NMR relaxation rates at the sites are affected via an effective correlation time $\tau_{eff,k}$, which depends on the lifetime in the site.

When the system presents chemical exchange as described in the former case, it is said that the system is in the exchange regime on the chemical shift time scale. Within this regime there are subdivisions, as it will be described in the next section. The case when the exchange affects the correlation times of the site is known as ultrafast exchange, and presents several different cases, depending on whether the strength of the interactions is the same in the different sites k [Wen]. Along this work we will be concerned **only with chemical exchange on the chemical shift time scale**.

The typical case of chemical exchange in NMR is provided by the two-site model in which a nuclear spin exchanges between sites A and B as described by

$$A \underset{k_B}{\overset{k_A}{\rightleftharpoons}} B \quad (3.4)$$

This kinetic scheme is of general relevance because a number of more complicated processes can be treated by defining pseudo-first-order rate constants. For instance, the case of using nuclear relaxation techniques in studying the way in which various ions affect protein conformation, and the way in which certain small molecules interact with macromolecules, is normally reduced to situations where there are two sites only, *bound* and *free*, with unequal populations and unequal relaxation times [CR]. The study of transverse relaxation patterns in tissues, with chemical exchange between extra- and intracellular water magnetization reservoirs is another example [BR].

In 1934, Erlemeyer and Gartner [EG] found, by deuterium labelling, that hydrogen exchanges quickly and completely between hydrogen peroxide and water. In general, the rate of exchange of hydrogen atoms bonded to oxygen is generally too fast to be measured by isotope labelling methods.

In 1958 Anbar et. al have studied the proton exchange between hydrogen peroxide and water [ALM] and found out that the inverse of the exchange rate (an effective exchange time) falls into the chemical shift time scale. In this chapter it is shown how this exchange can be used to quantify hydrogen peroxide concentrations in equilibrium aqueous solutions, as well as during a heterogeneous catalytic decomposition.

3.1.2 Theoretical Treatment of Two-Site Chemical Exchange

A more detailed description of two-site chemical exchange will be presented here. The effect of the exchange process depends on the nature of the NMR experiment. In this section we will be concerned with NMR Spectroscopy and transverse Relaxation using a Carr-Purcell-Meiboom-Gill pulse sequence (CPMG), although in our present work we have dealt only with experiments of the latter type. During the following treatment we will consider **free precession of two uncoupled spins subjected to exchange between two sites**, as shown schematically in (3.4), with a chemical rate constant given by [PKL],

$$k_{ex} = k_A + k_B = \frac{k_A}{p_B} = \frac{k_B}{p_A} \tag{3.5}$$

where p_A and p_B denote the equilibrium populations of spins in sites A and B, with $p_A + p_B = 1$. In his paper in 1991, Sobol remarked that the definition of average exchange rate in this way is non-standard, but he kept it to be consistent with the former people working on that. For first order reactions the quantities k_A and k_B are first-order rate constants. In other cases they are pseudo first-order rate constants. To make the treatment more general we will assume that the sites have, in the absence of exchange, different resonance frequencies in the rotating frame ω_A and ω_B in units of radians per second, being $\Delta\omega \equiv \omega_A - \omega_B$ their difference, as well as different relaxation times T_{1A}, T_{1B}, T_{2A}, T_{2B}. The only restriction is the assumption that $p_A \geq p_B$. Formulas appropriated for the opposite circumstances

3.1. Chemical Exchange in NMR

can be obtained simply by interchanging the labels. The equilibrium longitudinal magnetization at each site is $M_A^0 = p_A M^0$ and $M_B^0 = p_B M^0$.

Considering only the random process shown in (3.4), the evolution of the populations is governed by

$$dp_A/dt = -k_A p_A + k_B p_B$$
$$dp_B/dt = k_A p_A - k_B p_B \qquad (3.6)$$

The most evident effects of chemical exchange in NMR spectra are in the line shapes of the affected resonances. The evolution of the transverse magnetization subject to free precession under the Zeeman Hamiltonian is described by the Bloch equations, modified by McConnell [McC] to take into account the exchange process (as shown in eq. (3.6))

$$\frac{dM_A^+(t)}{dt} = -i\omega_A M_A^+(t) - \frac{M_A^+(t)}{T_{2A}} - p_B k_{ex} M_A^+(t) + p_A k_{ex} M_B^+(t)$$
$$\frac{dM_B^+(t)}{dt} = -i\omega_B M_B^+(t) - \frac{M_B^+(t)}{T_{2B}} - p_A k_{ex} M_B^+(t) + p_B k_{ex} M_A^+(t) \qquad (3.7)$$

where $M_A^\pm$ and $M_B^\pm$ denote the transverse magnetization in both sites, with $M_A^\pm = M_A^x \pm i M_A^y$ and $M_B^\pm = M_B^x \pm i M_B^y$ respectively. This system of equations is usually written in matrix form as,

$$\frac{d}{dt}\begin{pmatrix} M_A^+(t) \\ M_B^+(t) \end{pmatrix} = \begin{pmatrix} -i\omega_A - \frac{1}{T_{2A}} - p_B k_{ex} & p_A k_{ex} \\ p_B k_{ex} & -i\omega_B - \frac{1}{T_{2B}} - p_A k_{ex} \end{pmatrix} \begin{pmatrix} M_A^+(t) \\ M_B^+(t) \end{pmatrix}$$
$$(3.8)$$

Making the following definitions,

$$\alpha_A = \frac{1}{T_{2A}} + p_B k_{ex} + i\omega_A$$
$$\alpha_B = \frac{1}{T_{2B}} + p_A k_{ex} + i\omega_B, \qquad (3.9)$$

the system of differential equations can be written as

$$\frac{d}{dt}\begin{pmatrix} M_A^+(t) \\ M_B^+(t) \end{pmatrix} = \begin{pmatrix} -\alpha_A & p_A k_{ex} \\ p_B k_{ex} & -\alpha_B \end{pmatrix} \begin{pmatrix} M_A^+(t) \\ M_B^+(t) \end{pmatrix} \qquad (3.10)$$

Solutions of the system (3.10) are easily obtained using standard methods [AG1, Woc1, Sob1], with the form

$$\begin{pmatrix} M_A^+(t) \\ M_B^+(t) \end{pmatrix} = \mathbf{A}(t) \begin{pmatrix} M_A^+(0) \\ M_B^+(0) \end{pmatrix} \qquad (3.11)$$

where the matrix $\mathbf{A}$ consists of four time dependent elements,

$$\mathbf{A}(t) = \begin{pmatrix} a_{AA}(t) & a_{AB}(t) \\ a_{BA}(t) & a_{BB}(t) \end{pmatrix} \tag{3.12}$$

in which [PKL],

$$a_{AA}(t) = \frac{1}{2}\left[\left(1 - \frac{\alpha_A - \alpha_B}{\lambda_+ - \lambda_-}\right)\exp(-\lambda_- t) + \left(1 + \frac{\alpha_A - \alpha_B}{\lambda_+ - \lambda_-}\right)\exp(-\lambda_+ t)\right]$$

$$a_{BB}(t) = \frac{1}{2}\left[\left(1 + \frac{\alpha_A - \alpha_B}{\lambda_+ - \lambda_-}\right)\exp(-\lambda_- t) + \left(1 - \frac{\alpha_A - \alpha_B}{\lambda_+ - \lambda_-}\right)\exp(-\lambda_+ t)\right]$$

$$a_{AB}(t) = \frac{p_A k_{ex}}{\lambda_+ - \lambda_-}[\exp(-\lambda_- t) - \exp(\lambda_+ t)]$$

$$a_{BA}(t) = \frac{p_B k_{ex}}{\lambda_+ - \lambda_-}[\exp(-\lambda_- t) - \exp(\lambda_+ t)]$$

and

$$\lambda_\pm = \frac{1}{2}\left[\alpha_A + \alpha_B \pm \sqrt{[(\alpha_A - \alpha_B)^2 + 4p_A p_B k_{ex}^2]}\right] \tag{3.13}$$

The signal after a single pulse (FID) is $S(t) = M_A^+(t) + M_B^+(t)$, and the corresponding spectrum is given by its Fourier transformation. A few simulated spectra are shown in Fig. 3.1 for both symmetric and non-symmetric situations, i.e. symmetric exchange in which $p_A = p_B = 0.5$ (Fig. 3.1A) and asymmetric exchange in which $p_A = 0.7$ and $p_B = 0.3$ (Fig. 3.1B). In both cases the simulations were performed at 200 MHz Larmor frequency, with $T_{2A} = T_{2B} = 2\,s$, with the difference in the resonance frequencies being $\Delta\omega = 8.1 \times 10^{-3}\,rad\,s^{-1}$ (the reason for the choice of this particular value will become clear in the next section). The effect of exchange in the spectra is to shift the resonance positions and broaden the lines until the lines coalesce into a single line when $k_{ex} \approx |\Delta\omega|$. There is an important detail to notice here: there exists a discrepancy in units, but the coalescence arises when k_{ex} in Hz is equal to $\Delta\omega$ in units of $rad\,s^{-1}$, due to the fact that the Bloch-McConnell equations are written in the rotating frame. A more exact analysis of the eqs. (3.11)-(3.13) gives for the coalescence condition $2\sqrt{p_A p_B}k_{ex} \approx \Delta\omega$ [Woe3]. In both cases presented in Fig. 3.1 that corresponds to $k_{ex}/\Delta\omega = 1$ and $k_{ex}/\Delta\omega = 1.09$ respectively. As exchange become faster, a single resonance line is observed at the population-weighted average shift, $\omega_{av} = p_A\omega_A + p_B\omega_B$.

For equal populations the single line is centered at $\omega_{av} = 0$, while in the unequal

3.1. Chemical Exchange in NMR

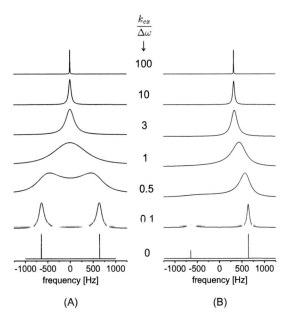

Figure 3.1: Spectra simulated for a system presenting two-site chemical exchange, with (A) equal populations ($p_A = p_B = 0.5$) and (B) skewed populations ($p_A = 0.7$, $p_B = 0.3$), as function of the quantity $k_{ex}/\Delta\omega$. The frequency difference, in the absence of exchange, was set to be $\Delta\omega/2\pi = 1290$ Hz, in a field strength of 200 MHz, for both cases. The transverse relaxation times were chosen to be $T_{2A} = T_{2B} = 2\,s$.

populations shown here, the corresponding frequency is $\omega_{av} = 1620\ rad\,s^{-1}$ (or 258 Hz).

As the exchange continues to increase, the resonance line shape becomes increasingly narrow until in the limit $k_{ex} \to \infty$, the relaxation time is given by $T_{2av}^{-1} = p_A T_{2A}^{-1} + p_B T_{2B}^{-1}$. The relative values of k_{ex} and $\Delta\omega$ define the limits of what is commonly referred to as **fast** and **slow** exchange on the **chemical shift time scale**:

- $k_{ex} <| \Delta\omega |$ Slow exchange

- $k_{ex} \approx | \Delta\omega |$ Intermediate exchange

- $k_{ex} > |\Delta\omega|$ Fast exchange

In the slow exchange limit $k_{ex} \ll \Delta\omega$, the relaxation times observed in both sites are related with their values in absence of exchange in the following manner:

$$\frac{1}{T_{2A}^{obs}} = \frac{1}{T_{2A}} + p_B k_{ex}$$

$$\frac{1}{T_{2B}^{obs}} = \frac{1}{T_{2B}} + p_A k_{ex} \qquad (3.14)$$

In the fast exchange limit, when $k_{ex} \gg \Delta\omega$, the relaxation time of the averaged resonance becomes

$$\frac{1}{T_{2av}} = \frac{p_A}{T_{2A}} + \frac{p_B}{T_{2B}} + \frac{p_A p_B \Delta\omega^2}{k_{ex}} \qquad (3.15)$$

If the populations are similar, then slow exchange is recognized easily by the presence of two resolved lines, while fast exchange is recognized by the presence of a single averaged resonance. However, if the populations are skewed, the minor component is preferentially broadened in comparison to the major component. Thus, in the slow exchange limit, the resonance at ω_B is both lower in intensity by a factor p_B/p_A and significantly broader by a factor $(T_{2B}^{-1} + p_A k_{ex})/(T_{2A}^{-1} + p_B k_{ex})$ than the resonance at frequency ω_A. As a result, if $p_A \gg p_B$, then the resonance at ω_B may be undetectable [Woe2]. Thus, as emphasized by Ishima *et al.*, the mere observation of a single exchange-broadened resonance does not necessarily indicate that the exchange process is fast on the chemical shift time scale [IT]. In Fig. 3.2 two simulated spectra are shown in order to observe that effect. Both spectra have been simulated with populations $p_A = 0.95$ and $p_B = 0.05$, with the same parameters employed in Fig. 3.1, for (A) $k_{ex}/\Delta\omega = 3$ and (B) $k_{ex}/\Delta\omega = 0.5$. Notice that the spectra are similar, with the second peak being unobservable in (B).

If the site populations are unequal with $p_A > 0.7$, Millet and co-workers have shown that the time scale for the chemical exchange can be determined from the static magnetic field dependence of exchange line broadening for the observable resonance, even if the resonance for the minor site cannot be detected in the slow exchange regime [MLK$^+$]. In general, the exchange broadening is defined by the excess contribution, T_{2ex}^{-1}, to the transverse relaxation rate, T_2^{-1},

$$\frac{1}{T_{2ex}} = \frac{1}{T_2} - \frac{1}{T_2'} \qquad (3.16)$$

where T_2' represents the relaxation time in absence of exchange. In slow exchange, the right-hand side of eq. (3.16) refers to site A, while in fast exchange it refers

3.1. Chemical Exchange in NMR

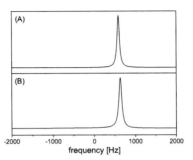

Figure 3.2: Spectra simulated in order to show the impossibility of observing the differences between slow and fast exchange when $p_A \gg p_B$. The same values than those used in Fig. 3.1 were employed here, except for the population which were chosen to be $p_A = 0.95$ and $p_B = 0.05$. The two cases correspond to (A) fast exchange with $k_{ex}/\Delta\omega = 3$; (B) slow exchange with $k_{ex}/\Delta\omega = 0.5$.

to the population-averaged resonance. For small changes in the magnetic field, the fractional change in the chemical exchange broadening, $\delta T_{2ex}^{-1}/T_{2ex}^{-1}$ and the fractional change in the static field, $\delta B_0/B_0$, are related by,

$$\frac{\delta T_{2ex}^{-1}}{T_{2ex}^{-1}} = \Upsilon \frac{\delta B_0}{B_0} \quad (3.17)$$

The constant of proportionality, or scaling factor Υ is defined by

$$\Upsilon = \frac{d(\log(T_{2ex}^{-1}))}{d(\log(\Delta\omega))} \quad (3.18)$$

For instance, in the limit $p_A \to 1$, and $T_{2A}^{-1} = T_{2B}^{-1}$, eqn. (3.18) yields [MLK+]

$$\Upsilon = \frac{2(k_{ex}/\Delta\omega)}{1 + (k_{ex}/\Delta\omega)^2} \quad (3.19)$$

Numerical calculations indicate that $0 \leq \Upsilon \leq 2$ provided that $p_A > 0.7$. Thus, Υ defines the NMR chemical shift time scale:

- $0 \leq \Upsilon < 1$ Slow exchange
- $\Upsilon \approx 1$ Intermediate exchange

Chapter 3. Chemical Exchange and Relaxation in H_2O_2

- $1 < \Upsilon \leq 2$ Fast exchange

It means that in the case of unequal populations with $p_A > 0.7$, by means of two independent experiments at different static magnetic fields (B_0), observing the broadening of the major peak, it is possible to compute Υ and determine the exchange regime.

The exchange averaging in high-resolution NMR spectra of fine structure produced by chemical shifts and/or the electron coupling of the nuclear spins has been used extensively for the study of fast rate processes [Mei, GH, GLM, LM1]. In some more complicated systems, rate studied have been made by calculating the entire line shape as a function of exchange rate and comparing the theoretical curves with those observed [AG1]. The analysis of the experimental data have been based generally upon a treatment based on the Bloch-McConnell equations, as shown above. Also, density matrix formulations have evolved for such studies [GVW].

As an alternative to line-width measurements of T_2, the pulse method of Carr-Purcell-Meiboom-Gill (CPMG) can be used [MG2], which is often more precise. In such a pulse sequence, a train of echoes is obtained by extending the $90° - 180°$ two pulse sequence with additional $180°$ pulses spaced at intervals of t_E, commonly called *echo time*, after the first $180°$ pulse, where the echo time is twice the separation between the $90°$ and the first $180°$ pulse. The successive echoes are formed in between the train of $180°$ pulses, and their amplitudes are modulated typically by an exponential decay (or a weighted sum of such decays). In Fig. 3.3 the sequence can be seen schematically.

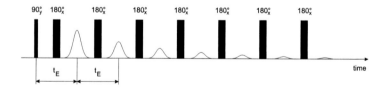

Figure 3.3: CPMG pulse sequence. The separation between the successive $180°$ pulses, or equivalently, the separation between echoes, is the so-called echo time t_E. It is twice the separation between the $90°$ and the first $180°$ pulse.

In non-exchanging systems, assuming a perfectly homogeneous static magnetic

3.1. Chemical Exchange in NMR

field, the particular choice of t_E has no effect on the value of T_2 obtained and it is therefore selected on the basis of experimental convenience. In the following treatment we will assume the perfect homogeneity of B_0. We will shift to the next section the discussion of the effect of inhomogeneous fields, and the convenience of choosing short echo times will become clear. Therefore, we are in the situation in which the echo time does not play any role in the determination of relaxation times in system in the absence of exchange. However, if the nuclei involved are exchanging between two different sites, under appropriate experimental conditions the observed value of T_2 from CPMG experiment will exhibit a variation with the value of t_E used.

When a nucleus jumps from one site of type A to a site of type B, the nuclear precessional frequency changes from ω_A to ω_B, and the dephasing rate changes correspondingly for that nucleus. When the CPMG pulse separation t_E is small enough, compared to the mean exchange lifetime, $\tau = k_{ex}^{-1}$, the irreversible dephasing resulting from the exchange is negligible and the observed T_2 value is [Jen1]:

$$\lim_{t_E \to 0} T_2^{-1} = \frac{(T_{2A} + T_{2B})}{2T_{2A}T_{2B}} + \frac{k_{ex}}{2} - \left[\left(\frac{(T_{2A} + T_{2B})}{2T_{2A}T_{2B}} + \frac{k_{ex}}{2}(p_B - p_A) \right)^2 + k_{ex}^2 p_A p_B \right]^{1/2} \tag{3.20}$$

This result does not involve the chemical shift.

On the other hand, if t_E is large compared to $\tau = k_{ex}^{-1}$, all the nuclei will have exchanged many times between two pulses, thereby undergoing the maximum additional irreversible dephasing so that the total observed T_2 is a lower limit [Jen1]:

$$\lim_{t_E \to \infty} T_2^{-1} = \frac{(T_{2A} + T_{2B})}{2T_{2A}T_{2B}} + \frac{k_{ex}}{2} - \Gamma \tag{3.21}$$

where

$$\Gamma = \Re \left[\left(\frac{(T_{2A} + T_{2B})}{2T_{2A}T_{2B}} + \frac{k_{ex}}{2}(p_B - p_A) \right)^2 - \frac{\Delta\omega^2}{4} + k_{ex}^2 p_A p_B \right.$$
$$\left. + i\Delta\omega \left(\frac{(T_{2A} + T_{2B})}{2T_{2A}T_{2B}} + \frac{k_{ex}}{2}(p_B - p_A) \right) \right]^{1/2}$$

Such a change in the transverse relaxation time as function of the pulse separation is usually referred to as *relaxation dispersion* or *Luz-Meiboom dispersion* [Sob2]. This problem has been largely studied and there exists abundant literature. In all the cases the problem is treated from the abovementioned solutions of the Bloch-McConnell equations for one pulse (eqn. (3.11)). Luz and Meiboom first predicted that T_2 measured with a train of echoes in a system governed by exchange depends on the echo separation [LM2]. However, they have presented a derivation with many restrictive assumptions. Allerhand and Gutowsky have extended the model, providing an elegant solution focused on closed formulas for relaxation times that were amenable to manual calculations [AG1, AG2]. In order to obtain such formulas, they assumed a system with two sites having the same inherent T_2, i.e. $T_{2A} = T_{2B} = T_2^0$. The most important result in their work is that the echo envelope decay will be *non-exponential*, but for most circumstances of experimental interest the decay is essentially exponential. Few years later, Carver and Richards have removed the limitation, providing closed formulas for the observed relaxation time, in the general form [CR]:

$$T_2 = T_2(T_{2A}, T_{2B}, k_{ex}, p_A, p_B, \Delta\omega, t_E) \tag{3.22}$$

They also found out, numerically evaluating their model, that for most experimentally accessible systems the decay is exponential.

The general solution is based on that presented by Allerhand and Gutowsky. The key resides in recognizing that the effect of a 180° pulse in the CPMG sequence is to invert the sense of precession of the nuclear spins. Employing the vectorial notation for the solution after a pulse (eqn. (3.11))

$$\mathbf{M}^+(t) = \mathbf{A}(t)\mathbf{M}^+(0) \tag{3.23}$$

where $\mathbf{A}$ is the matrix defined in (3.12) and

$$\mathbf{M}^+ = \begin{pmatrix} M_A^+ \\ M_B^+ \end{pmatrix}$$

the signal at different times, expressed as $S(t) = M_A^+(t) + M_B^+(t)$ can be calculated for the times marked in Fig. 3.4 by means of:

$\boxed{1}$ $\mathbf{M}^+(0)$

$\boxed{2}$ $\mathbf{M}^+(t_E/2) = \mathbf{A}(t_E/2)\mathbf{M}^+(0)$

$\boxed{3}$ $\mathbf{M}^+(t_E/2) = \mathbf{A}^*(t_E/2)\mathbf{M}^-(0)$

$\boxed{4}$ $\mathbf{M}^+(t_E) = \mathbf{A}(t_E/2)\mathbf{A}^*(t_E/2)\mathbf{M}^-(0)$

$\boxed{5}$ $\mathbf{M}^+(3t_E/2) = \mathbf{A}(t_E/2)\mathbf{A}(t_E/2)\mathbf{A}^*(t_E/2)\mathbf{M}^-(0)$

$\boxed{6}$ $\mathbf{M}^+(3t_E/2) = \mathbf{A}^*(t_E/2)\mathbf{A}^*(t_E/2)\mathbf{A}(t_E/2)\mathbf{M}^+(0)$

3.1. Chemical Exchange in NMR

$\boxed{7}$ $M^+(2t_E) = A(t_E/2)A^*(t_E/2)A^*(t_E/2)A(t_E/2)M^+(0)$

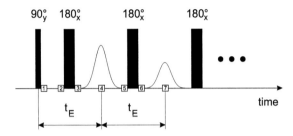

Figure 3.4: CPMG sequence only for the first two echoes. The numbers are related with the points successively calculated by expanding the solution of Bloch-McConnell equations for a single pulse.

In that way, solving recursively they have arrive to the close formulas for the echoes envelope decay. Although they have arrived to a bi-exponential solution we reproduce here only one of the T_2 expressions, that which they have remarked as describing the most experimentally accessible systems. Then,

$$T_2 = -t_E/\ln(\Lambda_1) \tag{3.24}$$

where Λ_1 is a complicated function of all the exchange parameters:

$$\ln(\Lambda_1) = -\frac{t_E\,\sigma_+}{2} + \ln\left[\left(D_+\cosh^2(\xi_+) - D_-\cos^2(\eta_-)\right)^{1/2} + \left(D_+\sinh^2(\xi_-) + D_-\sin^2(\eta_+)\right)^{1/2}\right]$$

with the complementary definitions:

$$\begin{aligned}
2D_\pm &= \pm 1 + 2(\ ^2 + 2\Delta\omega^2)/(\ ^2 + \zeta^2)^{1/2} \\
\xi_\pm &= (t_E/4\sqrt{2})\{\pm[+\ +(\ ^2+\zeta^2)^{1/2}]^{1/2}\} \\
\eta_\pm &= (t_E/4\sqrt{2})\{\pm[-\ +(\ ^2+\zeta^2)^{1/2}]^{1/2}\} \\
&= \sigma_-^2 - \Delta\omega^2 + 4k_A k_B \\
\eta &= 2\Delta\omega\sigma_- \\
\sigma_- &= (1/T_{2A} - 1/T_{2B} + k_A - k_B) \\
\sigma_+ &= (1/T_{2A} + 1/T_{2B} + k_A + k_B)
\end{aligned}$$

Chapter 3. Chemical Exchange and Relaxation in H_2O_2

In independent works, J. Jen has also encountered that the envelope of the CPMG train has a nearly exponential behavior, except for the first few echoes, first for the two-site case [Jen1], later extended to the more general multi-site exchange case [Jen2].

In Fig. 3.5 simulated curves based on eqn. (3.24) are presented, in order to show the T_2 dependence on t_E, for two different examples. In both cases it was assumed $T_{2A} = T_{2B} = 2\,s$ and $\Delta\omega/2\pi = 1290$ Hz. In Fig. 3.5A, the curves have been simulated with $p_A = 0.7$; different curves correspond to different values of $k_{ex}/\Delta\omega$. In Fig. 3.5B, the relative value $k_{ex}/\Delta\omega$ was set to be 0.1 and different curves were obtained varying p_A. It can be clearly seen how the T_2 value decays continuously between the two limiting values presented in eqns. (3.20)-(3.21). Notice that, for fixed populations (case (A)) even in slow exchange ($k_{ex}/\Delta\omega = 0.1$) T_2 changes about two orders of magnitude. On the other hand, at slow exchange (case (B)) a little amount of the substance B, i.e. p_A changing from 1 to 0.95, also produces a change about two orders of magnitude in the relaxation time. As a result, T_2 is very sensitive to changes in population as well as exchange rate in a CPMG experiment.

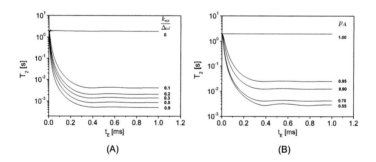

Figure 3.5: T_2 vs t_E simulated for different situations. (A) $p_A = 0.7$ and $k_{ex}/\Delta\omega$ varying from 0 to 0.9. (B) $k_{ex}/\Delta\omega = 0.1$ with p_A ranging from 0.55 to 1. In both cases $T_{2A} = T_{2B} = 2\,s$ and $\Delta\omega/2\pi = 1290$ Hz.

3.2 Chemical Exchange in Aqueous Peroxide Solutions

3.2.1 Previous Works

As mentioned above, aqueous hydrogen peroxide solutions present proton exchange between H_2O_2 and H_2O [ALM], and can be represented schematically by means of the general scheme shown in eqn. (3.4), associating the site A with water molecules and B with hydrogen peroxide molecules. Stephenson et al. have reported measurements where both peaks could be observed separately at concentrations below 0.2 % v/v at 400 MHz Larmor frequency. The area of the hydrogen peroxide peak in the slow exchange regime, and the dependence of the single peak position on H_2O_2 concentration in the fast exchange case were used by the authors to follow a homogeneously catalyzed hydrogen peroxide decomposition [3D]. This method presents the necessity of knowing the exchange regime of the sample at any time during calibration and reaction. That requirement represents a weakness due to the fact that, as mentioned in the last section, in the slow exchange regime and at low concentrations, the peak corresponding to protons in the H_2O_2 molecule is both lower in intensity and broader than the peak produced by protons in water. If $p_A \gg p_B$, the resonance at ω_B may be undetectable, making it very difficult to discriminate between fast and slow exchange. In addition to the problem at small concentrations, the method cannot be implemented in the case of heterogeneous catalysis due to the field disturbances produced by the presence of a porous medium containing metal.

On the other hand, the T_2 dependence on echo time in CP (Carr-Purcell) and CPMG experiments has been largely used in a wide range of applications:

- In the determination of the number of water molecules involved in the proton-transfer between trimethylammonium ion and trimethylamine in aqueous solutions [LM2].

- To investigate the hindered internal rotation of N,N-dimethyltrichloroacetamide (DMTCA) and dimethylcarbamyl chloride (DMCC) as a function of temperature [AG1].

- To determine the temperature dependence of the proton exchange time in water enriched at 4 % ^{17}O [KP].

- To study the relaxation patterns in tissues and relaxation mechanisms in protein solutions [BR, ZGA].

- In biomolecular dynamics, in order to characterize microsecond to millisecond dynamics in recognition and catalysis, [Akk, PKL].

The main goal of this work is to make use of this property to monitor the reaction evolution during the heterogeneous catalyzed hydrogen peroxide decomposition (eqn. (1.1)).

3.2.2 T_2 vs. echo time in Bulk Samples

In 3.1.2 the echoes envelope in a CPMG experiment performed in a liquid presenting exchange has been described. For the sake of clarity, it was supposed a perfect homogeneous static magnetic field. However, in a more realistic situation, the field must be considered inhomogeneous. Due to translational Brownian diffusion the spins move in space, and in presence of inhomogeneities, experience a time-dependent magnetic field. Because their Larmor frequency is slightly different in the intervals between pulses, the refocusing of partial magnetization is not perfect at times nt_E and the amplitudes of the echoes is lower than expected by solely relaxation mechanisms. Then, the transverse relaxation rate (the inverse of the relaxation time) extracted from a CPMG experiment, in a liquid sample, consists in an addition of two terms, $1/T_2 + f_D$. The former term contains all the information about the relaxation mechanisms themselves while the latter includes the effects of an inhomogeneous static magnetic field:

$$f_D = \frac{1}{12}\,\gamma\,g^2 D\,t_E^2 \qquad (3.25)$$

where g is the average local gradient field, and D is the self-diffusion coefficient of the liquid. In a general manner, it is useful to write the full equation for T_2 with all the variables involved in a generalized way, in the case of a liquid sample presenting two-site exchange,

$$T_2^{-1} = T_2^{-1}(T_{2A}, T_{2B}, k_{ex}, p_A, p_B, \Delta\omega, t_E) + \frac{1}{12}\,\gamma\,g^2 D\,t_E^2 \qquad (3.26)$$

At this stage it is convenient to accommodate the variables according to our problem. The relative populations, as mentioned, are related by $p_A + p_B = 1$. On the other hand, associating the hydrogen peroxide molecules with the site B, we can write

3.2. Chemical Exchange in Aqueous Hydrogen Peroxide Solutions 53

the concentration of H_2O_2 in water as $C = p_B$. Then, the above equation can be written,

$$T_2^{-1} = T_2^{-1}(T_{2A}, T_{2B}, k_{ex}, C, \Delta\omega, t_E) + \frac{1}{12}\gamma\, g^2 D\, t_E^2 \qquad (3.27)$$

In absence of exchange (in water, for instance) T_2 is independent of t_E as can be seen from the curves corresponding either to $k_{ex}/\Delta\omega = 0$ in Fig. 3.5A or $p_A = 1$ in Fig. 3.5B. Then, performing CPMG experiments for different echo times, the average deviation from a constant T_2 can be fully associated with the effect of molecules diffusing through a inhomogeneous field. On the other hand, in case of samples presenting exchange, both terms are echo time dependent and the effects are superimposed (eqn. (3.27)).

In order to observe the T_2 dependence on t_E in aqueous hydrogen peroxide solutions, a series of experiments was performed in bulk samples, for different concentrations of H_2O_2 diluted in water but keeping the total amount of liquid as well as the sample position relative to the resonator coil unchanged, to avoid unwanted variations in magnetic susceptibilities. The samples consisted of a 7 mm inner diameter tube filled with 1.1 ml of liquid. The different bulk samples were obtained by mixing distilled water (CHRO-MASOLV Plus, Sigma-Aldrich), with hydrogen peroxide 30 % by weight (Riedel-deHaen). The measurements were performed on a Bruker DSX200 spectrometer at a proton Larmor frequency of 200 MHz equipped with a 10 mm ID r.f. coil. The temperature of the experiments was set to 19°C.

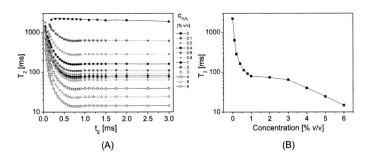

Figure 3.6: Transverse relaxation time corresponding to bulk samples, for different concentrations of H_2O_2 diluted in water. (A) T_2 vs. t_E presenting the features expected in samples presenting exchange (see Fig. 3.5). (B) T_2 vs. concentration at fixed echo time, $t_E = 600\,\mu s$

Figure 3.6A shows T_2 vs t_E with concentrations being indicated in % v/v. While the curve corresponding to $C = 0$, i.e. pure water, shows a slight dependence on t_E as a consequence of the static field inhomogeneities (maximum change $\simeq$ 10 % in the t_E range shown), the transverse relaxation time shows a much more pronounced variation when t_E increases at fixed concentration. It clear results that any effect of inhomogeneities is minor compared to the effect of exchange. For every concentration the transverse relaxation time reaches a plateau at $t_E \sim 600\,\mu s$. Beyond that echo time, the maximum difference in T_2 for different concentrations is observed as well. It means that fixing $t_E > 600\,\mu s$ it is possible to differentiate concentrations from measuring T_2. In theory every value is equivalent, but shorter values are preferred in order to reduce the effect of diffusion (see eqn. (3.25)). Figure 3.6B shows the relaxation time values obtained at $t_E = 600\,\mu s$ extracted form Fig. 3.6A as function of concentration. Notice that even at small concentrations (0.1 % v/v) the effect of exchange is clearly distinguishable from the pure water case.

3.3 Monitoring a Catalyzed H_2O_2 Decomposition

3.3.1 Monitoring the Decomposition by means of T_2 Measurements

Before using the concentration dependence of T_2 presented in Fig. 3.6B to follow the evolution of a heterogeneous catalytic reaction, we must demonstrate that for the system under investigation (being representative for a heterogeneously catalyzed reaction and serving as a model for actual reactors), the sample peculiarities such as catalyst shape, metal content, internal magnetic field gradients and susceptibility effects, are minor so that it is viable to compare the values obtained during the reaction with those obtained in bulk samples.

The presence of a metal containing sample, i.e. the porous medium, disturbs the surrounding magnetic field and, consequentially, the relaxation time obtained from a CPMG experiment (eqn. (3.25)). That effect can be taken into account by replacing g by a different, effective value of magnetic field gradient g_s in eqn. (3.27). However, there is another effect, related with the reaction itself, which might affect T_2 by means of eqn. (3.27): the oxygen bubbles produced during the decomposition, and their effect on the diffusion coefficient. As it was described in the previous chapter, the effective diffusion coefficient [CCS] D_{eff} in $z-$direction becomes dependent on

3.3. Monitoring a Catalyzed H_2O_2 Decomposition

the reaction time, t_r. When the reaction starts it presents a value much larger than the bulk diffusion coefficient, decreasing while the reaction rate is becoming smaller. At the end of the decomposition process, when the reaction ceases, it corresponds to D_0, the diffusion coefficient of water in the presence of the catalyst.

Figure 3.7A shows the evolution of D^z_{eff}/D_0 during the reaction in the vicinity of the catalyst pellet, measured in the conditions explained in the previous chapter, continuously every 4 minutes for a period of 7 hours, the initial concentration of H_2O_2 being 5 % v/v.

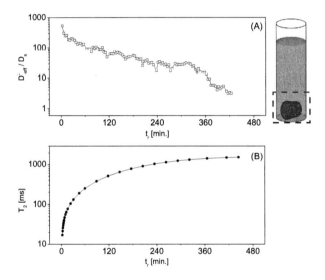

Figure 3.7: Effective diffusion coefficient and T_2 measured during decomposition in a solution of 5 % v/v H_2O_2 initial concentration, as function of reaction time. (A) Evolution of D^z_{eff}/D_0 during the reaction, measured in a defined closed volume in the vicinity of the catalyst pellet, continuously every 4 minutes for a period of 7 hours. (B) T_2 during the reaction, obtained from CPMG train with $t_E = 600\,\mu s$. (the experiment was performed independently of the previous one). Although the values were obtained every 40 seconds, only a few points are shown for the sake of clarity.

Figure 3.7B presents the evolution of T_2 during the reaction, for an experiment performed independently of the previous one, monitored every 40 seconds for a period of 7 hours by means a CPMG train with $t_E = 600\,\mu s$ (only a few points are shown for the sake of clarity). The setup is shown in the right-hand side of the figure, with the rectangle representing the coil's sensitive volume. It is important to point out that there is no signal originating from inside the pellet because T_2 is too short. In our experiments T_2 inside the pellet is shorter than 600 μs, the echo time used in the CPMG train, so that a contribution can be safely neglected in the data analysis which excludes the first echoes.

Notice that the higher values of D^z_{eff} correspond to the lower values of T_2. Considering the previous observations, the transverse relaxation time as a function of the reaction time t_r yields:

$$T_2^{-1}(t_r) = T_2^{-1}(T_{2A}, T_{2B}, k_{ex}(t_r), C(t_r), \Delta\omega, t_E) + \frac{1}{12}\,\gamma\,g_s^2 D^z_{eff}(t_r)\,t_E^2 \qquad (3.28)$$

In this equation, g_s was assumed to be time independent. Although the effective value of the magnetic field gradient depends on the oxygen bubbles produced by the reaction, and thus on the reaction time, the effect is much smaller compared with the magnetic field gradients generated by the presence of metal within the porous medium itself (which is time independent), so that it can be neglected. As the reaction proceeds and consequently D^z_{eff} decreases, the effect of the second term in eqn. (3.28) becomes smaller. At the same time, C decreases continuously, resulting in an increase of T_2 (see Fig. 3.6), corresponding to a decrease of the first term. In other words, the stronger effect due to enhanced diffusion occurs at the beginning of the reaction, when the transverse relaxation time is smaller (then the term T_2^{-1} due to only relaxation is larger); the decrease of the influence of the diffusion effect takes place at a later stage when the T_2 due to relaxation is larger (then T_2^{-1} smaller). A compensation in the relative effect of the second respect to the first term in eqn. (3.28) is expected.

In order to determine how much the catalytic reaction differs from the bulk results (considering the abovementioned), the following experiment was performed: a set of identical liquid samples was prepared with 5 % v/v hydrogen peroxide concentration, and the same number of pellets was selected randomly. The reaction in every tube was started at different times, labelled as $t_{ni}, t_{(n-1)i}, ..., t_{1i}$ and the measurement was performed at times $t_{nf}, t_{(n-1)f}, ..., t_{1f}$, so each sample has reacted a period of time $t_r = t_{kf} - t_{ki}$, respectively (see Fig. 3.8). The measurement of each sample consisted of two parts: firstly, the tube with the pellets inside was

3.3. Monitoring a Catalyzed H_2O_2 Decomposition

placed into the magnet with the reaction taking place, and T_2 was measured with a CPMG pulse sequence of a pulse separation $t_E = 600\,\mu s$. Immediately following this first measurement (duration $\simeq 40$ seconds) the pellet was removed from the tube and the liquid was shaken until complete removal of the bubbles was achieved, thus reproducing the identical conditions of the previously measured bulk samples. For these samples, T_2 was measured exactly with the same parameters. It is supposed here that all pellets are equivalent from the reaction point of view. This assumption was tested, measuring relaxation times for three different samples taken randomly, and the results between them were found to deviate by less than 2%. The plot in Fig. 3.8 shows both experiments for the complete set of samples. A difference between experiments can indeed be observed in every point, being more pronounced for longer t_r or equivalently at lower concentrations, but always being less than 15%. Such effect is expected, because at the end of the reaction, the relaxation time must be that corresponding to the tube filled with water plus the pellet, slightly different from the value obtained having only the liquid (T_2 is about 2 s but the effect of the metal will be present in the result).

From the plot it can be concluded that diffusion effects on the measured relaxation times are generally negligible, so that T_2 is dominated by proton exchange during the heterogeneously catalyzed decomposition.

On the other hand, temperature is an important variable in liquids experiencing exchange. Although from Fig. 3.8 it is clear that T_2 remains almost unchanged when comparing the bulk measurements with those performed at the beginning of the reaction (when the reaction rate is higher, leading to a maximum production of heat per unit of time), a simple test was made to estimate the effect. A tube with 5 % v/v initial concentration was placed into the magnet together with a thermocouple. Then a pellet was introduced, starting a reaction with T_2 and the temperature being monitored for a few minutes. The maximum change in temperature was observed to be $\sim 0.5°C$. Removing the pellet the temperature slowly decreased until reaching the initial value, T_2 presenting during this period a change $< 2\%$. This result confirms that the heat liberated during the reaction does not produce any observable change in the experiment.

In a further set of experiments, samples with different initial concentrations of hydrogen peroxide were measured. Three different tubes were prepared with the same set up as described above, i.e. identical tube sizes and amounts of liquid, with initial concentrations of 1, 3 and 5 % v/v, respectively. The relaxation time was

58 Chapter 3. Chemical Exchange and Relaxation in H$_2$O$_2$

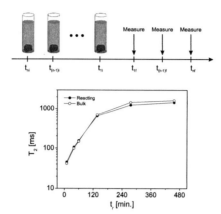

Figure 3.8: Experiment performed for testing the differences in T_2 during the heterogeneously catalyzed reaction from those values obtained in bulk samples. (TOP) The procedure is shown schematically. A set of tubes with different pellets was prepared and allowed to react for different periods. At certain time labelled as Measure, both values of T_2 were collected; one corresponding to the liquid reacting and the other one removing the pellet and stopping the reaction (i.e. equivalent to bulk samples). (BOTTOM) Both T_2 values plotted as function of reaction time.

monitored with the same parameters, (t_E and total number of echoes).

Figure 3.9A shows the evolution of T_2 for all three samples, during 11, 5.5 and 3 hours, respectively, corresponding to the initial concentrations of 5, 3 and 1 %, once more in the presence of an Al$_2$O$_3$ pellet containing Cu. The initial T_2 is shorter for higher concentrations, and grows smoothly towards the value of the transverse relaxation rate corresponding to pure water with the porous medium submerged, which corresponds to 2100 ms in these experiments. In Fig. 3.9B the results are presented in log − log scale in order to show that, in particular, T_2 initially follows the same dependence on reaction time for all three concentrations. As in the case of the experiments presented in Fig. 3.7 , data were acquired once every 40 seconds, but for the sake of clarity only a few points are plotted.

Determining the relaxation time is a fast and reliable method and can be carried out with a relatively good time resolution during a decomposition. Its smoothly and continuous change during the reaction makes it a suitable indicator for the on-line

3.3. Monitoring a Catalyzed H_2O_2 Decomposition

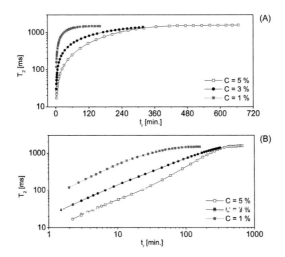

Figure 3.9: T_2 as function of reaction time for three different initial concentrations, in presence of an Al_2O_3 pellet containing Cu, in lin – log scale (A) and log – log scale (B). Data acquired once every 40 seconds. Only a few point are shown.

monitoring of the reaction evolution. However, a reliable quantification cannot be made only by considering the relaxation time. If the plot shown in Fig. 3.6B is used as a calibration curve and T_2 values obtained during the decomposition are transformed into H_2O_2 concentration vs. t_r, the results do not correspond to the expected values. Such a transformation is presented in Fig. 3.10: notice that at the beginning, when the concentrations are expected to be smaller than 5, 3 and 1 % respectively due to the lapse of time taken until the first point was acquired, the resulting concentrations correspond to 5.7, 4.5 and 0.55 respectively.

It is important to remark that all the experiments presented above, involving decomposition, have been performed with different pellets, starting always with the same initial condition: the pellets were randomly chosen from a batch, and they were always dry before the experiments. A completely different behavior was observed when either re-using the pellets or starting with different initial conditions (these two items might be related, depending on the case). Figure 3.11 shows the result of

Chapter 3. Chemical Exchange and Relaxation in H_2O_2

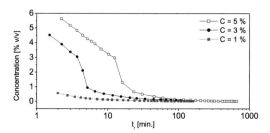

Figure 3.10: H_2O_2 concentration as function of t_r obtained transforming T_2 vs. t_r values shown in Fig. 3.9 by means of T_2 vs C curve of Fig. 3.6B.

seven different experiments using the same pellet, doped with Cu. In all cases the decomposition was monitored during one hour, with initial 5 % H_2O_2 concentration. The first curve was obtained starting with the sample in the same initial condition than in previous experiments, i.e. new sample completely dry.

From 2nd to 7th, the experiment was conducted in the following manner: after 1 hour of reaction, the pellet was removed from the liquid and submerged into a tube filled with water, where it was left for 20 minutes; then a new 5 % concentration solution was prepared and the reaction was started, representing the next experiment. In other words, the experiments are equivalent to those presented above, but using the same pellet, starting initially filled of water (except for the 1st).

In Fig. 3.11A the seven experiments are presented together to allow a better comparison. An apparent progressive decrease of dT_2/dt_r is observed from experiment to experiment. However, after the sixth realization the differences in slope become undetectable, and further experiments with the same pellet will yield the same result, within a narrow error interval. The plot in Fig. 3.11B shows only the first 25 minutes of the reaction. Notice that, at the very beginning after the 2nd realization, there is a decrease in T_2 which, in terms of Fig. 3.6B would represent an increase in the concentration. In a catalyzed H_2O_2 decomposition such increase is completely impossible. Moreover, the steady-state of the decomposition presents a succession increase-decrease-increase of T_2. In Fig. 3.11C T_2 vs. t_r is plotted in log −scale and then, at least for times longer than those where the initial oscillations are present, $d(\ln(T_2))/dt_r$ seems to be constant. From all the three plots it can be

3.3. Monitoring a Catalyzed H_2O_2 Decomposition

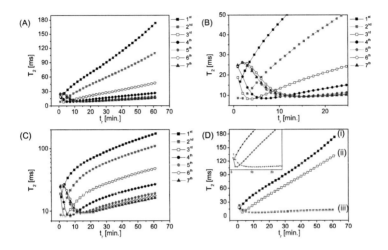

Figure 3.11: (A) Seven different equivalent experiments using the same pellet. (B) First 25 minutes of the reaction. (C) T_2 vs. t_r in log –scale. (D) Three equivalent experiments performed in the following conditions: (i) a new pellet initially dry, (ii) a new pellet pre-saturated with water and (iii) the same pellet used in (A)-(C), but two weeks later. The inset plot represents the first 25 minutes.

concluded that both the water present within the pellet at $t_r = 0$ and the successive use of the same sample considerably affect the results.

In order to decouple these effects more experiments were needed, monitoring the reaction of a pellet new and initially dry (i), a second pellet new but pre-saturated with water (ii), and the same pellet used 7 times before, but two weeks later (iii). The results are shown in Fig. 3.11D, where the inset plot represents the first 25 minutes. Comparing (i) and (ii) it can be concluded that the water within the pellet before the reaction starts is related with the decrease of T_2 at small t_r's. On the other hand, from (i) and (iii) (both initially dry) it is clear that when using the sample for decomposing H_2O_2, some irreversible changes occur within the porous medium.

Given the new behaviors observed, it was necessary to repeat the tedious experiment shown in Fig. 3.8, now employing a re-used pellet. In order to have a highly reliable results, 20 different pellets with their respective tubes were employed. All of them were put to react 8 times before the measurement, paying special attention in that they finish the 8th reaction just before starting the decomposition which was going to be measured. The whole preparation consumed more than 30 hours, finally resulting in the same experiment than that shown in Fig. 3.8 (and detailed in the text following the figure) with each pellet in the steady-state. The results are summarized in Fig. 3.12

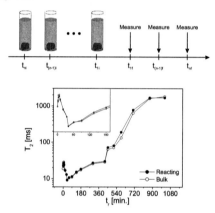

Figure 3.12: Experiment performed to see whether the pellet has influence in the T_2 oscillatory behavior at small t_r's. The procedure was analogous to that presented in Fig. 3.8, with an extra item: previous to every sample preparation, the pellets were used 8 times to decompose 5 % of H_2O_2. Both T_2, with the reaction going on and the corresponding bulk value obtained by quickly removing the pellet are plotted as function of reaction time. The inset plot show the first 3 hours in detail.

From the results it is clear that the decrease in relaxation time is independent of the presence of the catalyst particle. Despite some small differences, which might be associated with the fact that removing the pellet takes a time while the reaction continues, the bulk samples follow the same trend as the reacting samples. Notice that the first bulk measurement does not have any corresponding reacting point. That point was measured alone, to have a well defined starting T_2 value, and was associated to $t_r = 0$.

Once it was proven that it is still possible to compare the relaxation times during

3.3. Monitoring a Catalyzed H_2O_2 Decomposition 63

the reaction with the bulk values, longer experiment were carried out. Figure 3.13A shows 15 hours of T_2 evolution corresponding to the case presented in Fig. 3.11D (iii). Even when the decrease in T_2 represents a very small fraction of its total range of variation, transforming the data into concentration by means of the calibration curve (Fig. 3.6B) leads to an apparent increase in concentration of about 50 % (see Fig. 3.13B).

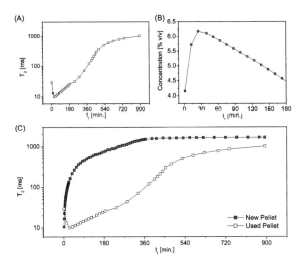

Figure 3.13: (A) T_2 vs reaction time during 15 hours of decomposition where a re-used pellet has been employed as catalysts. (B) H_2O_2 concentration as function of reaction time, obtained transforming the curve shown in (A) by means of Fig. 3.6B. (C) T_2 vs t_r during 15 hours from two experiments performed independently with a new pellet in one case and the pellet re-used in the other case.

Although the information given by monitoring the relaxation time cannot be directly translated into concentrations, some information about relative reaction rates can be extracted. In Fig. 3.13C the curves corresponding to 15 hours of decomposition are plotted, in the case of employing as catalyst either a new pellet or that pellet used many times before. For the reaction itself, it is expected to obtain only water when it has ceased. Therefore the final T_2 value must coincide independently of the catalyst. Based on that, from the plot it can be concluded that there is a reduction in the *catalytic activity* as the pellet is re-used; in the former case the liquid has almost reached the expected final relaxation time after 6 hours of reaction, while in the other case, even after 15 hours the value is perceptible below.

At this stage, no more information about the decomposition than just the total time taken for the liquid to reach the expected final T_2 value, can be extracted. It could be still useful, for example, to test the efficiency of catalyst particles impregnated with different amount of metals, different metals, sizes or even geometries. Nevertheless, it represent a global information, while the more detailed information related with the time evolution of the reaction remains unaccessible within the shape of the T_2 vs. t_r curves.

3.3.2 The pH as an Independent Quantity

In the previous treatment it was stated that, for the system under investigation, T_2 shows a remarkable dependence on relative populations, i.e. H_2O_2 concentration, and exchange rate (or its inverse, the average lifetime). It has been also pointed out that the temperature is another important parameter to control, in order to get reliable results. The sample temperature enters into the problem via the exchange rate, which is normally affected by changes in the thermal conditions. On the other hand, Anbar et al. have found, in their work on equilibrium reactions in aqueous hydrogen peroxide solutions, that pH is another variable which strongly affects the exchange rate [ALM]. They have reported a completely different behavior on the exchange rate depending on whether the solution is in the range of pH < 4 (*acidic*) or pH > 5 (*basic*). The values within 4 < pH< 5 represent a transition interval between basic and acid regions.

A set of bulk samples was prepared in order to explore the pH dependence on concentration. Figure 3.14A shows the results, obtained at room temperature, by means of a pH-meter equipped with a glass-body electrode, with single junction (Hanna Instruments). Notice that the pH values monotonically decreases when concentration increases. By means of the T_2 vs. C curve shown in Fig. 3.6B it can be transformed into a T_2 vs pH plot (Fig. 3.14B). In our experiments we have started with 5 % H_2O_2 conc. and finished with water. It means that during the decomposition the liquid *crosses* the transition region corresponding to 4 < pH < 5. However, no extra effects might be expected, *a priori*, when transforming the relaxation times

3.3. Monitoring a Catalyzed H_2O_2 Decomposition 65

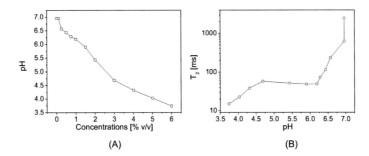

Figure 3.14: (A) pH vs. C obtained from a set of bulk samples at room temperature. (B) T_2 vs pH obtained from the combination of (A) and Fig. 3.6B.

into concentration using Fig. 3.6B. The argument can be summarized as follows. Given a liquid sample with certain concentration, it will have a T_2 value associated, but also a pH value. Now the liquid is decomposed for a certain period of time and the reaction is stopped. The new concentration will have another T_2 value but also the corresponding pH value. On the other hand, the full information is already contained in the plot of Fig. 3.6B. In other words, that plot can be seen now as (pH,T_2) vs. C plot rather than only a T_2 vs. C. Based on that, a transformation or translation of T_2 into Concentration would not produce any extra effect, as for example, the oscillation at the beginning of the reaction.

Nevertheless, although the previous argument sounds reasonable, its validity depends on the assumption that the reaction does not have any effect on the liquid's pH value, except for that accompanying the change in concentration. It is supposed that the reaction, just *transforms* the pair (pH,T_2) related to the initial concentration, into that pair corresponding to the final concentration. Or, equivalently, it supposes that every reaction should follow the path shown in 3.14B in a plot T_2 vs. pH with the time as a parametric variable.

There is a strong argument against that supposition: the complexity of the interaction between the liquid and the metal and metal oxides at the catalytic surface. The interaction, and then the products of the decomposition, depend on variables as type of metal, its concentration, different metal oxide species and their distribution, geometry of the pore space, and the mechanism of decomposition itself. As a result, it cannot be supposed that under a catalytic decomposition, any given H_2O_2 concentration is directly connected with an unique (pH,T_2) pair. In order to test that conclusion in our system, a simple and short experiment was performed: a tube with 5 % H_2O_2 bulk concentration was placed into the magnet and T_2 was monitored during 2 hours. When the equilibrium temperature was reached (due to differences between room and magnet temperature, it normally takes $\sim$ 1.5 h) the pH value was measured, without moving the tube from the magnet, giving pH $= 4.05 \pm 0.02$. Then the pellet was introduced for a period of 2 minutes, and removed quickly. The pH was measured once again, this time yielding pH$= 4.32 \pm 0.02$.

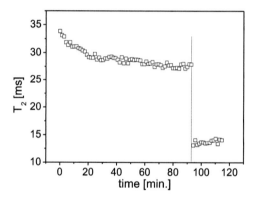

Figure 3.15: Experiment performed to observe the extra change in pH produced by the pellet. The vertical line indicates the reaction starting point. Previously, the bulk sample was allowed to reach thermal equilibrium inside the magnet. Then the pellet was introduced and the reaction started.

In Fig. 3.15 the results are plotted, with the reaction starting point marked. After it has been reacted 2 minutes, T_2 fell down from $27.2\,ms$ to $13.1\,ms$ with the corresponding increase of pH in ΔpH $= 0.27$. This clearly indicates that the presence of the pellet itself generates a change in pH and T_2 which does not have any relation with those values expected from Fig. 3.14B. It is important to remark that the difference in T_2 during the first 1.5 h is related with the difference in $\sim 5°$ between the laboratory and the magnet. The pellet reacting changed the temperature by

3.3. Monitoring a Catalyzed H_2O_2 Decomposition

~ 0.5°, but 20 minutes after it was removed, the thermal equilibrium was reached again with no appreciable change in T_2, as can be seen from the right part of the plot.

At this stage, and due to the fact that all the variables mentioned above, some of them possible to control (type of metal and, somehow, concentration) but others impossible to quantify (as porous geometry or distribution of metal oxides), it is convenient to attack the problem in a more general way. It was decided to treat the concentration and pH as independent variables. It is possible to observe experimentally the influence of both variables in T_2 for bulk samples, considering

$$T_2 = T_2(T_{2A}, T_{2B}, k_{ex}(\text{pH}), C, \Delta\omega, t_E) \qquad (3.29)$$

Varying the relative populations and pH of the samples, and keeping all the other variables fixed, it would be possible to have a 2-D experiment rendering the surface $T_2 = T_2(\text{pH}, C)$. Such surface would represent all the possible H_2O_2 states. In a reaction condition, the global effect of all the uncontrolled variables will arise as the manner in which they connect the independent variables accessible for monitoring, as pH and T_2, resulting in a much more reliable on-line quantification of the H_2O_2 concentrations.

In order to carry out the experiment in a highly precise manner, the H_2O_2 concentration of the reservoir was determined by means of titration with standarized $KMnO_4$, and each μ-pipet used for preparing the bulk samples was calibrated in at least 3 different volumes. By means of a thermocouple it was checked that even under wide differences in room temperature the sample inside the magnet was always at 19°C. The whole experiment consisted on the following steps:

(1) A given concentration was prepared by carefully mixing the necessary amounts of H_2O_2 and pure water.

(2) The tube with the sample was then placed into the magnet.

(3) During a period of time between 1 and 2 hours T_2 was monitored in order to observe with high precision, when exactly the sample was in thermal equilibrium. Once in equilibrium, T_2 was measured for different echo times.

(4) The sample was removed from the magnet, and pH was determined.

(5) A small drop (typically a few μl) of either H_2SO_4 or NaCl at different concentrations was added to decrease or increase the pH value respectively. The tube was placed back into the magnet. The procedure was done as fast as possible in order to avoid temperature changes of the sample.

Chapter 3. Chemical Exchange and Relaxation in H_2O_2

(6) The steps (3)-(5) were repeated, now consuming less time in the step (3), depending on how fast the steps (4) and (5) could be done. The repetitions were made until covering, for a given C, pH values ranging from ~ 3.5 to ~ 10

(7) The process (1)-(6) was repeated for different concentrations.

It must be remarked that the pH meter was calibrated periodically during the experimental time (typically every approx. 5 hours). Different concentrations of the base and the acid were used in order to limit the number of drops, keeping then the H_2O_2 concentration within 2 % of the original value. All the samples consisted of 1.75 ml of solution.

In that way a set of 120 samples were measured. Figure 3.16A shows T_2 vs. pH for **concentrations = 0 - 0.1 - 0.25 - 0.5 - 0.75 - 1 - 2 - 3 - 4 - 5 - 6 % v/v**. Notice that for water ($C = 0$) there is no change as pH changes, as expected. A considerably small concentration of H_2O_2 (0.1 %), on the contrary, is enough to produce an appreciable dependence on pH. The experiments were not continued beyond $C = 6$ % because at that concentration, the difference with, for example $C = 5$ %, was about 6 ms (in the whole pH range until the curves coalesce), and it was expected to be closer for the other curves, thus experimentally difficult to differentiate.

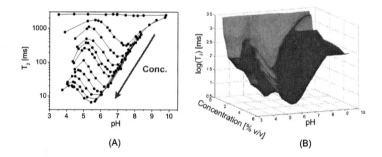

Figure 3.16: Results of the T_2 and pH measurements performed in bulk samples at different concentrations. (A) T_2 vs. pH for curves at constant concentrations: 0.1, 0.25, 0.5, 0.75, 1, 2, 3, 4, 5 and 6 % v/v. The curve which does not present change corresponds to water, i.e. $C = 0$. (B) A surface obtained from (A) by performing a 2D interpolation. That surface represents all the possible (C, pH, T_2) states of H_2O_2.

For the concentrations covered in that experiments, the curves present an oscil-

3.3. Monitoring a Catalyzed H_2O_2 Decomposition

latory behavior, reaching the maximum T_2 value in the interval $4.5 < \text{pH} < 5.5$. It follows a decay after which all of them coalesce in a unique curve. That curve acts as a "barrier", separating the *possible* (pH,T_2)-states on the left from the *forbidden* (pH,T_2)-states on the right side. A few points measured with pH < 4.5 suggest that a similar behavior might be expected in case of extending the range to lower values (except perhaps for the point corresponding to $C = 0.1\%$).

The plot shows all the possible (pH,T_2)-states at the particular temperature set for the experiments. In case of having a bulk sample with unknown concentration, one can measure pH and T_2, placing then the point into the plot. An H_2O_2 concentration can be assigned to the sample by choosing that C-curve which includes the point. Figure 3.16B shows a surface obtained by means of a 2-D interpolation (with a piecewise cubic interpolation). The accuracy in determining a concentration depends on the local shape of the curves at the particular (pH,T_2) values. In our experiments T_2 was acquired with high precision, checked in many cases by repeating the measurement several times and calculating the standard deviation. We can safely assume that, for the case of bulk measurements, the error in T_2 is always less than 3 %. The pH measurements, on the other hand, depend on the device used. In the experiments performed here, the measurements were really stable, with typical errors of ± 0.02. Nevertheless, that corresponds to the inherent error of the apparatus, but usually there are other error sources. By measuring several times different samples in a wide range of pH values, it was found out that it is safer to assume an accuracy of ± 0.05.

With those estimations, the error in C can be extracted directly from the curves presented in Fig. 3.16. In general, for C lower than 1 % v/v, the accuracy is rather high from pH=5 until the curves become indistinguishable. For instance, the point corresponding to the pair $(\text{pH}, T_2) = (5.5, 275\ ms)$ gives $C = (0.75 \pm 0.02)\%$ v/v. On the other hand, the pair $(\text{pH}, T_2) = (5.15, 16.7\ ms)$ gives $C = (4 \pm 0.1)\%$ v/v. The closer the point measured to the "barrier", the lower the accuracy. Within the barrier, it is impossible to distinguish between different concentrations.

All what has been said here lays on the assumption that the temperature is the same for every point in the curves. In the bulk experiments much care was taken to fulfill the condition, due to the fact that different positions in the surface have different "sensibility" with respect to changes in temperature.

3.3.3 Monitoring the Reaction by means of Simultaneous pH and T_2 Measurements

In the discussion presented above, it was pointed out that the plots in Fig. 3.16 univocally represent all the possible states of aqueous hydrogen peroxide solutions. All the others inaccessible variables related with the catalyst itself can be studied, then, by observing the effect they produce in pH and T_2. In every single catalyzed decomposition, those "hidden" variables will lead to a different path in the space (pH,T_2). By means of the procedure mentioned above, those paths can be transformed into a C vs. t_r plot, allowing thus comparisons between different catalyst in terms of metal content, porosity, etc. In order to illustrate the method, in this section we will present such procedure applied to the Cu-catalyst previously described in this chapter. However, in view of the possibility of monitoring the reaction until concentrations much lower the 0.1 % v/v, some modifications were made in the setup. In the interpretation of Fig. 3.8 it was remarked that the final stages of the reaction will be affected by the presence of the pellet. While $T_2 \sim 2.3\ s$ in our bulk experiments, when measuring close to the catalyst particle, that value is $T_2 \sim 2.1\ s$. That difference of about 200 ms would lead for instance, in case of having pH = 7.2, to a minimum $C \sim 0.01$ % v/v, while for pH = 6, to $C \sim 0.02$. Although those values present a relatively high accuracy, it is possible to diminish the effect of the metal containing sample by a simple modification in the layout. In previous experiments, the sensitive volume of the coil was placed to be aligned with the center of the pellet. A new experiment was performed, consisting on moving the tube down (or equivalently the center of the sensitive volume up) step by step and comparing the relaxation times to the bulk value, in order to directly observe how much the pellet affects the results. Figure 3.17 shows the curves obtained, with a simple scheme in the right hand side of the plot (height of liquid and size of the pellet are not to scale).

The first curve (open circles) represents the T_2 values of bulk water with the middle of the coil's sensitive volume ($\sim 1.2\ cm^3$) aligned with the center of the virtual position of the pellet. The decay observed is, once more, due to the difference between the room and magnet temperature. Once the equilibrium was reached, the value had a dispersion of $\pm 10\ ms$. In that condition, the pellet was introduced, and several experiments were performed to be sure the temperature was stable. Then, step by step, the tube was moved down, 3 mm each step, and T_2 was monitored.

3.3. Monitoring a Catalyzed H_2O_2 Decomposition 71

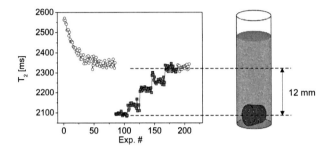

Figure 3.17: Plot showing the effect of the catalyst in T_2 at different heights relative to the coil's sensitive volume. When the distance is 12 mm, the relaxation time of water with and without the catalyst are similar.

The gray squares represent the values for every set of measurements. The last step corresponds to the center of the coil at 12 mm from the center of the pellet. Then the particle was carefully removed, and T_2 was monitored again to compare (open squares). The values agree within 10 ms, resulting in a minimum possible C to be obtained accurately about 0.001 % v/v for pH = 7.2 and 0.005 % v/v with pH = 6. It represents a factor of 10 and 4, respectively, compared to the minimum values possible to identify measuring closer to the catalyst particle. Notice that, in calculating the lowest C experimentally accessible, we were not concerned with temperature, but just used the T_2 accuracy. This absence is completely reasonable due the fact that at those concentrations the reaction rate is much lower than at the very beginning, so $\Delta T \ll 0.5°C$ can be assumed.

With the new setup, several experiments including reactions were made to test the accuracy with the bubbles disturbing the liquid. The experiments always consisted of allowing a given H_2O_2 concentration to reach the equilibrium temperature, and starting a reaction for circa 10 minutes. Then the T_2 was measured, the pellet removed smoothly, and a new T_2 value was acquired. The results (not shown) for all the combination of C and pH explored, were always below 5 %, i.e. $e_{T_2}/T_2 < 0.05$, where we have denoted with e_{T_2} the difference between the relaxation times with and without reaction.

After all the test concerning the accuracy, a reaction with simultaneously mon-

72   Chapter 3. Chemical Exchange and Relaxation in H_2O_2

itoring of pH and T_2 was planned. The experiment was carried out, as all the previous ones, in a Bruker DSX200 MHz device. Due to the impossibility of accessing to the tube with the pH-meter (mainly geometrical impediments but also perturbations of the electrodes produced by the high magnetic field), it was decided for the experiment to be conducted in the following manner:

(1) A tube containing 5 % v/v initial H_2O_2 concentration was placed into the magnet and T_2 monitored until the equilibrium temperature was reached.

(2) Then, the whole NMR probe containing the tube was quickly removed from the magnet, pH was measured (time estimated ~ 30 s), and the probe was returned to the magnet.

(3) A Cu-pellet hanging by means of a tiny wire was dropped into the tube, through the open bore of the magnet. Thus the reaction started.

(4) At certain t_r, T_2 was measured (exp. time 40 s) and the pellet removed by carefully pulling the wire.

(5) The whole probe was removed again and pH measured, as in (2).

(6) The steps (3)-(5) were repeated many times for different t_r covering a total experimental time of 15.5 hours. Proceeding in this way made it impossible to observe the evolution of the reaction with the so-called new pellets (see Fig. 3.11), so the pellet was pre-used several times before starting the experiment. Figure 3.18A shows the evolution of pH during the experiment, while in Fig. 3.18B, T_2 vs. t_r is plotted in log-scale. Notice the oscillation in relaxation times at the very beginning, in contrast with pH, which increases monotonically in the whole experimental time. These two plots can be combined to obtain, at any t_r, the pairs $(pH(t_r), T_2(t_r))$. Those pairs are plotted superimposed either onto the bulk $pH - T_2$ curves, in Fig. 3.18C or onto the interpolated surface in Fig. 3.18D. From the plot in (C) it can be seen that the initial point is placed just before the local maximum in the $C = 5$ % curve, so few minutes after the reaction has started, the "geometry" of the space forces T_2 to decrease. Notice that, in combination with the increase in pH during this lapse of time, it gives a decrease of the concentration. The effect is better observed in (D): the concentration always decreases even when the path falls into the "valley". The combined information was transformed into H_2O_2 concentration by assigning the value of the corresponding curve closest to the successive (pH, T_2) pairs, in the interpolated data. In order to obtain a measure for the accuracy, the error for every point was calculated numerically: the assignment of a concentration value was done not only to the pairs (pH, T_2) but also to $(pH \pm e_{pH}, T_2 \pm e_{T_2})$, where

3.3. Monitoring a Catalyzed H_2O_2 Decomposition 73

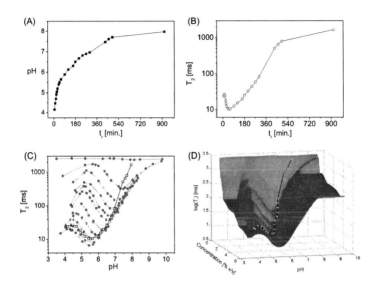

Figure 3.18: Results of an experiment where T_2 and pH were monitored simultaneously during 5 % of H_2O_2 decomposing in the presence of a pellet doped with Cu. (A) pH vs. t_r. (B) T_2 vs. t_r. (C) T_2 vs. pH plotted superimposed onto the bulk curves corresponding to different concentrations. (D) The information of T_2 and pH plotted as a path in the surface (C, pH, T_2)

e_{pH} was taken to be 0.05 and e_{T_2}/T_2, 5 %. Thus, the effect of the surface details in different regions when the data fluctuate can be taken into account. This procedure is similar to the standard error dispersion calculation, here made numerically.

Figure 3.19 presents the results of the transformation. In the left-hand plot, C vs. t_r is shown, with a pronounced fluctuation in C as well as its error, highlighted. The plot into the right-hand side is included in order to show in which region of the plane pH – T_2 the data pair is located. Those points are located near the barrier, where all the concentration curves coincide, making their differentiation impossible. About 2 hours later, the reaction evolution takes the path away from the barrier, and the accurate quantification of the concentration is possible again. In Fig. 3.20A, the same plot is shown, without including those points. Notice that the error bars

74   Chapter 3. Chemical Exchange and Relaxation in H_2O_2

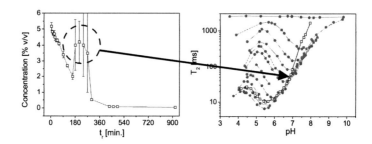

Figure 3.19: C v. t_r obtained from the information presented in Fig. 3.18. The concentration values were assigned by searching the bulk concentration curve closest to the successive (pH, T_2) pairs. The values with bigger errors correspond to the region in the bulk samples were it is impossible to distinguish concentrations.

are bigger at the beginning of the reaction, according with the proximity of the C-curves close to 5 % v/v. On the other hand, for low concentrations, the error bars are smaller than the size of the marker in the plot. Within the whole curve, however, the relative error follows $e_C/C < 5\%$. In Fig. 3.20B an equivalent curve is presented, with the transformation having been made using only the T_2 information. The oscillation in C is present at the beginning, leading to a fictitious increase on the concentration by a factor of 1.5 (from 4.25 to 6.5). By comparing both curves, it becomes clear that the inclusion of the pH information removes that undesired effect. In both plots, an inset is included with a zoom of the firs 3 hours of reaction.

3.4 Exchange Rates

In this section we demonstrate how information about the exchange rates can be obtained by means of the bulk experiments. In 3.3.2, a set of experiments performed in order to observe the effect on relaxation times produced by changes in pH for different concentrations was presented. It has been pointed out there that, for any single point for which a pair (pH,C) was determined in Fig. 3.16A, T_2 was measured for different echo times t_E, although the plot was constructed at fixed $t_E = 600\ \mu s$. Thus, the plot has an extra dimension not shown, t_E. In that dimension, every single point with determined (pH,C) has a shape similar to the curves shown in

3.4. Exchange Rates

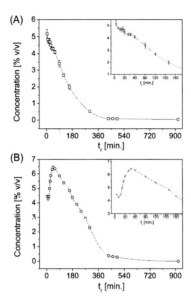

Figure 3.20: (A) C vs. t_r obtained by using the evolution of pH combined with T_2, once the points presenting the higher fluctuations were removed (see Fig. 3.19). (B) C vs. t_r using only the relaxation time information during the reaction. In both cases the inset plots show the first 3 hours of reaction in detail.

Fig. 3.6A. In eqn. (3.29), on the other hand, we have remarked that $k_{ex} = k_{ex}(pH)$. Combining all the variables, and recalling the relationship between k_{ex} with k_B via eqn. (3.5), the relaxation time measured can be written dependent on,

$$T_2 = T_2(T_{2A}, T_{2B}, k_B(pH), C, \Delta\omega, t_E) \tag{3.30}$$

At given concentration and pH, the dependence of T_2 on t_E can be fitted making use of the model presented in eqn. (3.24). The relaxation time of water, T_{2A} was extracted from the curve corresponding to $C = 0$. On the other hand, in 1958, Anbar et al. [ALM] have estimated the peak separation between protons in H_2O_2 and water in absence of exchange, i.e $\Delta\omega$, in about 6.3 ppm. At the Larmor frequency they have worked, it had been impossible to observe both peaks in the spectra. In 2005, Stephenson et al. [SB] could observe both beaks separately at 400 MHz Larmor frequency, encountering for the difference, $\Delta\omega = 6.4$ ppm. Although they have not reported the error for their result, based on the data shown in their work, we have estimated the error to be ± 0.05 ppm. The remaining variables can be taken as independent, and it is possible to find the best values by permitting them to vary within a finite range, choosing the value which minimize the difference between calculated curves and the experimental results. The free variables here are T_{2B} and k_B. The fittings with two variables, however, are usually time consuming, many times giving more than one result. In order to avoid that multiplicity and save calculation time, a different strategy was adopted. For few concentrations covering the whole range (0.1-6 % v/v) and for a collection of pH values, the best value of k_B and T_{2B} was found numerically. Then T_{2B} was allowed to change in a much wider range while keeping k_B fixed. It was encountered that once fixed T_{2A}, T_{2B} can be varied from a small fraction to a value 3 times bigger of T_{2A} without appreciable change in k_B. That behavior might be associated with the relatively low concentrations treated here. So, $T_{2B} = T_{2A}$ was assumed for simplicity, and the model was used with only one free variable, k_B.

Figure 3.21 shows the result obtained from the fittings. In Fig. 3.21A, the lifetimes of the protons in an H_2O_2 molecule, τ_B, is plotted vs. pH for five different concentrations: $C = 0.1, 0.5, 1, 3, 5$ % v/v. Although the fittings gave satisfactory results for all the other concentrations, only few were included for the sake of clarity. Notice that the lifetimes are of the order of milliseconds for $C = 0.1$ % and decrease as the concentration increases. Figure 3.21B shows the rate at which the protons leave the H_2O_2 molecules, $k_B = 1/\tau_B$, as function of pH. The plot shows both regions, according to the results presented in Anbar's work [ALM]. For pH > 5.5 the rate depends on the concentration, while for pH < 4.5, not so evident here due to the range of pH covered, the rate is concentration independent. The order of magnitude of the rates coincide with Anbar's results as well, although a direct comparison is not possible due the fact that their experiments were conducted at different temperature (27°). It has been decided to present here the dependence of k_B upon pH to directly compare it with the results obtained by Anbar et al. Nevertheless, similar plots can be obtained for the exchange rate, transforming back k_B into k_{ex} by means of eqn. (3.5). Recalling the equation, we see that $k_{ex} = k_B/p_A$, and due to the small concentrations used here (thus $p_A \sim 1$), we can expect the same global features for a k_{ex} vs pH plot.

3.4. Exchange Rates

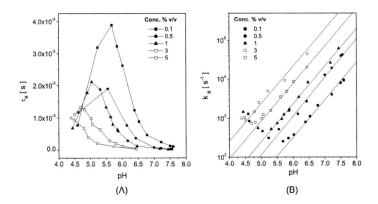

Figure 3.21: Result obtained from the fits of every single point of Fig. 3.16A in the dimension not shown, t_E by means of the model presented in eqn. (3.24). (A) The lifetime of the protons in a H_2O_2 molecule, τ_B, plotted vs. pH for five different concentrations: $C = 0.1, 0.5, 1, 3, 5$ % v/v. (B) $k_B = 1/\tau_B$, as function of pH, plotted to shows both regions corresponding to the results presented in Anbar's work [ALM].

The lifetimes, or equivalently the rates, are independent of the Larmor frequency of the experiments. They represent a property of the liquid at given temperature, concentration and pH. On the other hand, the effect of this exchange on T_2 depends on the frequency at which the experiments are performed, via eqn. (3.30). So, the values obtained here can be used to feed the model once more, and calculate the dependence of relaxation time on t_E for different frequencies. In Fig. 3.22 curves simulated for different values of pH and Larmor frequencies are presented. On the left-hand side, T_2 vs. t_E is shown for different concentrations, at 200 MHz Larmor frequency (as used in the experiments). Three different values of pH were used, 4.5, 5 and 6. The same concentrations and pH were used to simulate the curves at 40 MHz Larmor frequency. That particular frequency was chosen as being representative for commercial desk devices. The plots are presented on the right-hand side of the figure.

From the plots corresponding to pH=4.5 it can be observed that, at both frequencies the concentrations are distinguishable, although the higher the concentration the smaller the T_2 difference, as discussed before. For pH=5, the curves corresponding to $C = 0.5$ and $C = 0.75$ are almost indistinguishable. For pH=6, at 200 MHz the splitting is maximum between different concentration curves, making this range the best to quantify low concentrations. Nevertheless, between concentrations 4 and 6 % v/v the difference is on the order of a few milliseconds. On the contrary, at 40 MHz it is possible to clearly distinguish concentrations in the range 0.1 to 1 %. The curves corresponding to $C = 2-4$ are rather close, and the curves for $C = 5$ and 6 % have longer T_2 than the smaller concentrations. This overlapping is new compared to the experiments at higher frequencies, and it makes impossible to associate univocally the relaxation times with concentrations. Finally, notice that for all pH values, the maximum differences are encountered beyond $t_E = 600$ μs at high frequency, while at low frequency the maximum splitting is beyond $t_E = 1.5$ ms, although with $t_E = 1$ ms the splitting is acceptable for a relatively high accuracy.

It can be concluded that, in principle, it is possible to extend the method presented here to desktop spectrometers operating at lower frequencies, resulting in a much simpler experimental setup. In such devices it could be much easier to monitor pH and T_2 continuously during the reaction, without the necessity of removing the probe in each pH measurement. The temperature control is another advantage in performing the experiments without taking the sample out of the magnet. There are, however, some disadvantages in lowering the frequency, related with the maximum C possible to determine accurately and the slightly longer t_E needed, which could make the relaxation experiments more vulnerable to the enhanced diffusion.

3.4. Exchange Rates

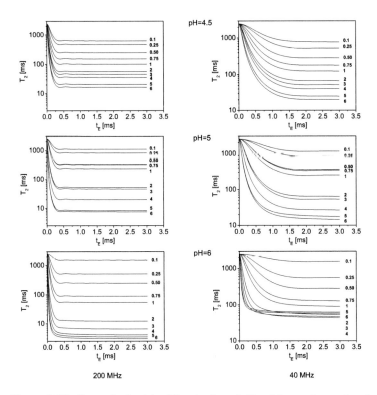

Figure 3.22: Curves simulated for different values of pH and Larmor frequencies. On the left-hand side, T_2 vs. t_E is shown for different concentrations, at 200 MHz Larmor frequency. On the right-hand side, the same simulations are presented at 40 MHz Larmor frequency. Three different values of pH were used, 4.5, 5 and 6.

Chapter 4

H$_2$O$_2$ Reaction Evolution Studied by NMR Imaging

Nothing is invented and perfected at the same time
Latin Proverb

4.1 Introduction

NMR Imaging (NMRI) is a noninvasive analytical technique, which is capable of producing images of arbitrarily oriented slices through optically nontransparent objects [Blu]. In 1973, Lauterbur reported the first reconstruction of a proton spin density map using NMR [Lau], and, in the same year Mansfield and Grannell [MG1] independently demonstrated the Fourier relationship between the spin density and the signal acquired in the presence of a magnetic field gradient. Since that time the field has advanced rapidly to the point where NMRI has a spatial resolution as good as that of the unaided human eye [Cal]. By far the dominating application of NMR Imaging is found in the medical field where it has become a routine method, and the number of clinical tomographs nearly saturates the market in most industrialized countries. NMRI of inanimate objects takes place in quality control and some special applications, but is still most strongly represented in academic research [Sta].

NMR Imaging technique has added a third dimension to the NMR-spectroscopy and NMR-relaxation phenomena. Therefore, NMR as practiced today can be represented by three partially overlapping areas representing spectroscopy, relaxation and imaging studies, as shown in Fig. 4.1 [Fuk].

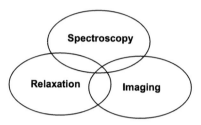

Figure 4.1: A schematic representation of NMR as partially overlapping areas representing the three main categories.

In this chapter we will be concerned with experiments which shall be placed in the region where Imaging and Relaxation overlap in the scheme shown in Fig. 4.1.

In an NMR imaging experiment, the image brightness is a direct measure of the NMR signal intensity obtained under the particular pulse sequence used in the image acquisition. This brightness does not only reflect spin density, but is weighted by the chemical constitution, the hardness or softness of the material, the diffusion coefficient and the translational motion or motion on a much faster time scale, depending on the design of the sequence. The contrast of an image reflects the heterogeneity of the sample with respect to the chemical and physical properties mentioned. This is because these properties reflect on NMR parameters such as NMR signal amplitude, frequency, T_1 and T_2 relaxation times and line shape. Therefore, the NMR sequence can be designed accordingly to obtain the desired contrast. The wide range of contrasts that can be imposed on a spin-density image is a unique power of NMR imaging [SH].

The wide variation of T_2 in an H_2O_2 decomposition appears to be an ideal factor to use as a contrast in NMRI. Monitoring the image intensity as the reaction advances would allow to have a global idea about the reaction stage, simultaneously in different parts within the catalyst particle. Along this chapter we will describe the experimental setup as well as the different tests performed in a spherical-like porous medium particle containing Pd as catalyst sites. The experiments presented here represent a first step in the long way leading to an online spatially resolved monitoring of a catalytic decomposition in fixed-bed reactor.

4.2 NMR Imaging

4.2.1 Basics

With the increasing popularity of NMR Imaging, in the last two decades many books were published presenting the principles of the technique. The variety in the level of description and technical applications is large, according with the wide range of problems in which it can be used as an invaluable tool. Based on the existence of such abundant literature, it was considered that a detailed exposition of the rudiments in this work would result redundant. Therefore, just a simple description of the use of field gradients in NMR Imaging is going to be presented here, strictly focused on the pulse sequence used in our experiment, the Spin Echo pulse sequence. More details concerning to that particular pulse sequence and its variants, as well as to other different techniques are discussed in many books. Particular examples are: *Principles of Nuclear Magnetic Resonance Microscopy* (Paul Callaghan, [Cal]), *NMR Imaging of Materials* (Bernhard Blümich, [Blu]), and *Principles of Magnetic Resonance Imaging* (Zhi-Pei Liang and Paul C. Lauterbur, [LL]).

In the imaging experiments, the magnetic field gradient is used in three different ways to yield spatial distribution of the spins. The first is to create an FID or a Spin Echo, in the presence of a magnetic field gradient, called **read-out** gradient, so that the signal contains the spatial information as a distribution of Larmor frequencies. The spatial distribution of the spins along the read-out gradient axis is extracted by a Fourier transformation of the temporal signal.

The second is to turn on a magnetic field gradient after the initial excitation but turn it off for the detection of the NMR signal. The Fourier transform of the signal will now be a spectrum of the nuclear spins without any information as to the spatial distribution of the spins except that the phase of the spectrum will reflect the contribution of the magnetic field gradient at the sites of the nuclei. Therefore, spatial information along such a **phase-encoding** gradient is obtainable if a series of experiments are performed with phase-encoding gradients of different magnitude, and a Fourier transform of the signals along the experiment coordinate is performed.

The third use of the magnetic field gradients is for slice selection, which is a method to selectively excite spins in a given spatial volume. Selective excitation is based on the concept that spins will absorb energy only at the Larmor frequency. Suppose we turn on a gradient, so that spins at different coordinates have different

84   Chapter 4. H$_2$O$_2$ Reaction Evolution Studied by NMR Imaging

Larmor frequencies. If we now excite the spins with a narrow band rf magnetic field, then only those spins whose Larmor frequency lie in this narrow band of frequencies will be affected, giving us an excited slice. The width of the slice can be controlled by either changing the frequency width of the excitation rf pulse or the value of the gradient. The location of the slice can be changed by shifting the center frequency of the excitation pulse. If the nuclear spin system being studied is a linear response system, the time modulation of the rf pulses need to be the Fourier transform of the frequency response which corresponds to the desired slice shape in the gradient [CF].

4.2.2   Spin Echo Pulse Sequence

We will be concerned here with a Spin Echo pulse sequence (SE) in 2 dimensions, with the echo signal being phase-encoded in one dimension (labelled as **phase**) and then acquired in the presence of a frequency-encoding gradient (labelled as **read**). The procedure is repeated using each time a different value for the amplitude of the phase-encoding gradient, resulting in a 2D data set. In the pulse sequence used here, the 180° pulse was set to be selective, with three sidelobes truncated sinc shape. In Fig. 4.2 such a sequence is presented with the relevant parameters. The three axes (read, phase, slice) can be aligned with any spatial direction, i.e. x, y or z to obtain a 2D image with slice selection. For the sake of clarity, and without loss of generality, we will assume an $x-y$ image with slice in z−direction. Although the choice might appear arbitrary, it corresponds to all the experiments presented in this chapter.

As shortly described in 4.2.1, frequency encoding makes the oscillation frequency of the signal linearly dependent on its spatial origin. If we consider an idealized one-dimensional object with spin distribution along the x−axis $\rho(x)$, and an applied magnetic field gradient g_r in that direction, the Larmor frequency at position x will be

$$\omega(x) = \omega_0 + \gamma g_r x \qquad (4.1)$$

Correspondingly, the signal generated locally from spins in an infinitesimal interval dx at position x, with the omission of the transverse relaxation effect is,

$$dS(x,t) = \rho(x)dx \exp[-i\gamma(B_0 + g_r x)t] \qquad (4.2)$$

The signal received from the entire object in the presence of this gradient is, in the

4.2. NMR Imaging

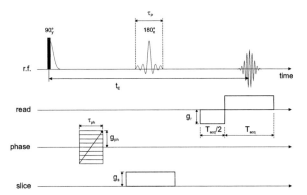

Figure 4.2: 2D Spin Echo pulse sequence used to acquire a 2D NMR Image. A 90° hard pulse is used to excite all the spins in the sample, while a 180° soft pulse is used to refocus only those spins within the desired slice. The signal is **first encoded** by a phase-encoding magnetic field gradient (g_{ph}) and then it is acquired in the **presence** of a magnetic field gradient in the read-out direction (g_r).

rotating frame

$$S(t) = \int dS(x,t) = \int_{-\infty}^{+\infty} \rho(x) \exp(-i\gamma g_r xt) dx \qquad (4.3)$$

By making the simple variable substitution,

$$k_x = \bar{\gamma} g_r t \qquad (4.4)$$

where $\bar{\gamma}$ denotes $\gamma/2\pi$, the following Fourier relationship is obtained,

$$S(k_x) = \int_{-\infty}^{+\infty} \rho(x) \exp(-i2\pi k_x x) dx \qquad (4.5)$$

It is clear that the role of the frequency-encoding gradient g_r is to map a time signal to a k–space signal.

On the other hand, if we consider a one-dimensional object with spin distribution along the y–axis $\rho(y)$, and an applied magnetic field gradient g_{ph} in that direction during a time interval τ_{ph}, the signal from an infinitesimal interval dy at position y

will be,

$$dS(y,t) = \begin{cases} \rho(y)\exp[-i\gamma(B_0 + g_{ph}y)t] & \text{if } 0 \leq t \leq \tau_{ph} \\ \rho(y)\exp(-i\gamma B_0 t)\exp(-i\gamma g_{ph}\, y\, \tau_{ph}) & \text{if } \tau_{ph} \leq t \end{cases} \quad (4.6)$$

By using the time interval τ_{ph} as a preparation period, the signal collected afterwards (see Fig. 4.2) will have an initial phase angle,

$$\phi(y) = -\gamma g_{ph}\, y\, \tau_{ph} \quad (4.7)$$

Since the phase is linearly related to the signal location y, the name phase-encoding becomes clear. As above, the total signal from the object is, in the rotating frame

$$S(t) = \int dS(y,t) = \int_{-\infty}^{+\infty} \rho(y)\exp(-i\gamma g_{ph}\, y\, \tau_{ph})dy \quad (4.8)$$

which yields the Fourier relationship

$$S(k_y) = \int_{-\infty}^{+\infty} \rho(y)\exp(-i 2\pi k_y y)dx \quad (4.9)$$

where equivalently to eqn. (4.4), the following definition was adopted,

$$k_y = \gamma g_{ph}\tau_{ph} \quad (4.10)$$

The sinc shape pulse used here, in the presence of a rectangular field gradient g_s, is expected to excite only those spins sited within the interval of width Δz, the thickness of the desired slice. It is possible to demonstrate [LL] that using a sinc with n side lobes of total length t_P, Δz is given by

$$\Delta z = \frac{4n\pi}{\gamma g_s t_P} \quad (4.11)$$

The thinnest slice that can be selected is limited by the available gradient strength and the shortest possible pulse length. The center of the slice, z_0 can be displaced by moving the center of the r.f. frequency.

4.2.3 Two-Dimensional Imaging

Let's denote $I(x,y)$ the desired 2-dimensional image of a three dimensional object characterized by the spin density $\rho(x,y,z)$. In case of using the slice selective pulse

4.2. NMR Imaging

to excite spins only within the interval $(z_0 - \Delta z/2, z_0 + \Delta z/2)$, the function $I(x,y)$ is expressed as

$$I(x,y) = \int_{z_0-\Delta z/2}^{z_0+\Delta z/2} \rho(x,y,z)dz \qquad (4.12)$$

The signal to be acquired is, thus,

$$S(k_x, k_y) = \int_{-\infty}^{-\infty} \int_{-\infty}^{-\infty} I(x,y) \exp[-i2\pi(k_x x + k_y y)] dx dy \qquad (4.13)$$

Therefore, a basic task of a sequence to produce a 2D image is to generate sufficient data to cover the k-space. A further Fourier transformation of the data will give as a result the image $I(x,y)$. There exist many different strategies to fully sample the k-space (see for example [BKZ]).

To understand how k-space is traversed in the SE pulse sequence (Fig. 4.2), lets consider the n-th excitation, i.e. the n-th step in the phase-encoding gradient. Just after the phase-encoding period, we will have $k_x = 0$ and $k_y = \gamma g_{ph}^n \tau_{ph}$, leading to a vector in the k-space

$$\mathbf{k}_A = (0, \gamma g_{ph}^n \tau_{ph}) \qquad (4.14)$$

where g_{ph}^n denotes the value of the n-th step of the phase-encoding gradient. If we consider an experiment with N_{ph} points in phase, $g_{ph}^1 = -g_{ph}$ and the superscript runs until the last point corresponds to $g_{ph}^{N_{ph}} = g_{ph}$ (see Fig. 4.3).

The subsequent 180° pulse will invert the phase of the spins, giving a new vector $\mathbf{k}_B$ which follows

$$\mathbf{k}_B = -\mathbf{k}_A \qquad (4.15)$$

In the scheme shown in Fig. 4.3, the phase-encoding gradient was supposed to vary from $-g_{ph}$ to g_{ph} and $n < N_{ph}/2$, so $\mathbf{k}_A$ results negative, whereas $\mathbf{k}_B$ is positive.

The first part of the field gradient applied in read direction, of amplitude g_r and duration $T_{acq}/2$ produces a variation in k_x of $\gamma g_r t'$ with $0 < t' < T_{acq}/2$. At the end of the period, we will have the vector

$$\mathbf{k}_C = (-\gamma |g_r| T_{acq}/2, -\gamma g_{ph}^n \tau_{ph}) \qquad (4.16)$$

Then the last period of the read gradient start, leading to a evolution of k_x as $\gamma g_r(t - T_{acq}/2)$, so $\mathbf{k}_D$ yields,

$$\mathbf{k}_D = (\gamma |g_r| (t - T_{acq}/2), -\gamma g_{ph}^n \tau_{ph}) \qquad (4.17)$$

The signal acquisition will start simultaneously with the last period of the read gradient, and a spin echo is obtained at $t = T_{acq}/2$. The procedure is schematically shown in Fig. 4.3. Repeating it N_{ph} times for $n = 1, ..., N_{ph}$ the whole 2D k-space will be covered.

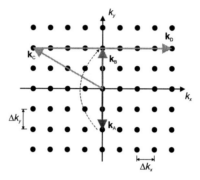

Figure 4.3: A 2D k-space schematically shown. The arrows show how it is traversed during a Spin Echo pulse sequence.

4.2.4 Sampling Requirements of k-space and FOV

The key to multidimensional imaging lies in generating a sufficient number of signals to cover the k-space. Although it represents a multidimensional problem, in practice, one treats sampling along each dimensions separately (as discussed above), thus reducing it to a one-dimensional sampling problem. Here we treat only the sampling requirements for a 2D image to be acquired with the sequence shown in Fig. 4.2.

Let's assume we have a 2-dimensional object bounded by a rectangle of width FOV_x and FOV_y as shown in Fig. 4.4A, where FOV stands for **Field Of View**, the maximum width of the window in the real space to be imaged. According to the sampling theorem (see for example [Mar]) we have

$$\Delta k_x \leq \frac{1}{FOV_x} \quad \text{and} \quad \Delta k_y \leq \frac{1}{FOV_y} \qquad (4.18)$$

known as the *Nyquist sampling criterion*, where Δk_x and Δk_y are the separation between successive k-space points in both direction respectively, as shown in Fig. 4.4B. As we have assumed that the frequency encoding is used in the x-direction and phase encoding is used in the y-direction, we have

$$\Delta k_x = \gamma |g_r| \Delta t$$
$$\Delta k_y = \gamma \Delta g_{ph} \tau_{ph} \qquad (4.19)$$

4.2. NMR Imaging

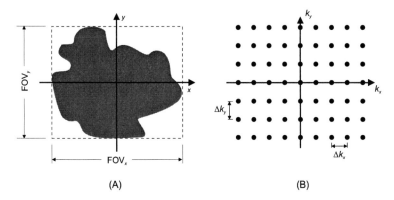

Figure 4.4: (A) A 2D object to be imaged, enclosed by a rectangle of dimensions FOV_x and FOV_y, where FOV stands for Field Of View. (B) Scheme of the k-space with the relevant parameters. The separation between points is related to the FOV in different directions by means of eqns. (4.18)

where Δt is the time separation between acquired points in read direction and Δg_{ph} is the phase-encoding gradient step.

Substituting eqn. (4.19) into eqn. (4.18), we immediately obtain the requirements in the data acquisition parameters:

$$\Delta t \leq \frac{1}{\gamma |g_r| FOV_x} \quad (4.20)$$

$$\Delta g_{ph} \leq \frac{1}{\gamma \tau_{ph} FOV_y} \quad (4.21)$$

They represent the relationship that the sampling interval time and the phase-encoding step height have to satisfy to avoid loss of image information when Fourier transforming the k-space data set obtained.

4.2.5 Contrasts and Pixel Intensity

As mentioned in 4.1, different pulse sequences produce different contrasts reflected in the pixel's brightness. More specifically, the pixel intensity is a multi-parameter function of spin density ρ, relaxation times T_1, T_2 and T_2^*, diffusion coefficient D, and so on. We may, then, express the image intensity as

$$I_p = f(\rho, T_1, T_2, T_2^*, D, ...) \tag{4.22}$$

where the exact functional form depends on the pulse sequence to be used.

Consider the pulse sequence of Fig. 4.2, and let's denote by M_z^0 the magnetization before the 90° pulse. If the waiting time between repetitions, t_R, is much longer than any T_1 present in the sample, the amplitude of the spin echo signal will be

$$A_E = M_z^0 \exp(-t_E/T_2) \tag{4.23}$$

On the other hand, if the repetition time is reduced (but holding $t_R \gg t_E$), the magnetization before the 90° pulse reaches a steady state M_z'

$$M_z' = M_z^0 [1 - \exp(-t_R/T_1)] \tag{4.24}$$

Therefore, the amplitude of the spin echo signal as function of t_E and t_R results

$$A_E = M_z^0 [1 - \exp(-t_R/T_1)] \exp(-t_E/T_2) \tag{4.25}$$

Thus, after the Fourier transformation, the pixel intensity carries information about spin density and transverse and longitudinal relaxation times simultaneously,

$$I_p \propto \rho [1 - \exp(-t_R/T_1)] \exp(-t_E/T_2) \tag{4.26}$$

However, we can selectively emphasize one of the contrast mechanisms by properly choosing the sequence parameters t_R and t_E. For instance, if a short t_E is used (with $t_E \ll T_2$), the term $\exp(-t_E/T_2)$ approaches unity and the T_2-weighting factor can then be ignored.

A very different view of the understanding of the contrast is to regard any extra dimension in addition to the mere image in a multi-dimensional imaging experiment as "contrast". For example, an extra dimension representing different t_R or t_E can be acquired. Then, the third dimension can be fitted with the proper function, pixel by pixel, to give T_1 or T_2 for every pixel. It is usually called a parameter image. We should also clarify the term weighted image and parameter image: a weighted image still contains the spin-density in the signal amplitude and represents a classical contrast image, whereas the parameter image only contains the pure parameter (i.e. velocity, diffusion coefficient, T_1, T_2) values in each pixel of the image.

4.3 Reaction Evolution Inside the Pd-Catalyst

4.3.1 The Sample

In the previous chapter it was pointed out that, for the cylindrical pellets used, T_2 was extremely short within the pore space. This shortening of the relaxation time arises due to the relaxivity of the porous walls itself, enhanced in that case by the presence of metal [KEM, FAK]. The corresponding values for the pellets doped with either Cu or Pt were in the range of 500-600 μs, much shorter than the minimum t_E accessible (see Fig. 4.2) with our device. Although there are methods suitable for imaging samples with such short (and even shorter) relaxation times, they do not allow the use of slice selection [Cal, Blu, LL]). In our experiments, on the other hand, due to the liquid surrounding the pellets the implementation of slice selection is mandatory.

In view of the limitations, the experiments were carried out for different kind of samples. In this case, we have opted for sphere-like pellets, from BASF (R0-20/72 PDE) with an outer shell containing 0.72 wt-% Pd as catalyst metal. The pellets have sizes between $2 - 4$ mm in diameter, but we have chosen those with larger diameter so that it was possible to maximize the number of pixels per image inside the pellet, due to the fact that the FOV is constant, corresponding to the inner diameter of the tube (in order to prevent aliasing effects). In Fig. 4.5 a picture of a whole pellet as well as a slice of it, in order to show the Pd layer, is shown. The picture is accompanied by an $x - y$ NMR image of the pellet submerged in water, with 7×7 mm and slice selection in $z-$direction of 500 μm thickness, with 32 points acquired in read direction and 32 steps in phase direction. The number of scans was set to be 1024, with $t_E = 1.85$ ms (the shortest possible), and the repetition time $t_R = 125$ ms yielding a total experimental time of 70 minutes. The resultant resolution is about 220 μm in both directions. The difference in brightness correspond to different pixel intensities, carrying information about T_1, T_2 and spin density, ρ (eqn. 4.26). Notice that, due to the large difference between t_R and water's T_1 in bulk, the signal corresponding to the water outside the pellet is considerably reduced, giving a good contrast with the signal from the water inside, whose T_1 is of the order of t_R.

As the relaxation times change during the reaction, it would be directly translated into changes in pixel intensity. It is the aim of this chapter to use the time

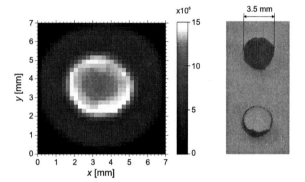

Figure 4.5: 2D $x - y$ image of a Pd pellet submerged in water, with slice selection in z−direction, obtained with a Spin Echo pulse sequence as shown in Fig. 4.2. The FOV in read as well as phase direction is 7 mm, and the slice thickness is 500 μm. The experiment was performed acquiring 32 points in read direction, with 32 steps in phase direction, $NS = 1024$ and $t_R = 125\ ms$. On the right-hand side a picture of the pellet is shown, with a cut to observe the Pd layer.

dependence of I_p as an indicator of the reaction progress within the porous medium.

4.3.2 Initial and Final States of the Reaction

The first step in using the time dependent relaxation parameters as contrast consisted on estimating the initial and final values expected during reaction conditions. In the experiments presented here, the decomposition was always performed starting with 5 % H_2O_2 concentration. In bulk, T_2 would vary from tenths of milliseconds to seconds. On the other hand, bulk T_1 experiments (data not shown) show that it increases from $3\,s$ to $3.2\,s$ as the H_2O_2 concentration increases from $C = 0$ to $C = 5$ % v/v, so that, in a decomposition it would decrease a few hundred of milliseconds. Nevertheless, the values are different inside the pellet, due to the interaction of the liquids with the pore walls, as was mentioned above. In order to quantify this effect, both relaxation times were measured inside the porous medium with spatial resolution.

In order to obtain the so called T_2-maps, i.e. a parameter image representing the T_2 value corresponding to every pixel, in terms of the nomenclature used in 4.2.5, a

4.3. Reaction Evolution Inside the Pd-Catalyst

series of images with different t_E was acquired, setting t_R longer than any T_1 present inside the pellet. In that case, the second factor in eqn. (4.26) is close to unity, yielding the pixel intensity proportional to ρ and T_2. However, in the discussion presented in 4.2.5 any effect of internal gradients produced by susceptibility differences between the liquid occupying the pore volume and the pore walls was omitted. As discussed in 3.2.2, the effects of the gradients on the signal can be taken into account by multiplying eqn. (4.26) with a factor $\exp(-1/12\,\gamma^2 g^2 D t_E^3)$ to eqn. (4.26). Thus, fitting I_p in the third dimension with a function of the form

$$I_p = I_0\,\exp(-t_E/T_2)\,\exp(-Bt_E^3) \qquad (4.27)$$

it is possible to obtain at the same time a T_2- and a $B-$map, where $B = 1/12\,\gamma^2 g^2 D$.

A T_2 map of the pellet submerged in water will give us the expected final situation, and another one submerged in 5 % H_2O_2 will represent the initial situation of the reaction. However, the map with hydrogen peroxide is impossible to obtain without having a reaction. In order to avoid the decomposition, two different pellets (with approximately the same size) were carefully *peeled* until the complete removal of the Pd from the surface was achieved. The reason of using two different pellets instead of one responds to the impossibility of completely removing the liquid from inside small pores once the pellet was saturated.

Figure 4.6 shows the T_2-map corresponding to the pellet saturated with water. The values cover the T_2 range $3 - 5.5\ ms$, presenting the larger values close to the particle's center.

The images for the T_2-map were acquired with 64 points in read direction and 32 steps in phase direction, with a FOV of $7 \times 7\ mm$ and the slice thickness 500 μs. The resulting resolution was about 110 μs and 220 μs in read and phase directions respectively. All the images were acquired with $NS = 32$ and $t_R = 1\,s$ leading to a total time for every image of circa 17 minutes. The echo time was varied from 1.85 to 15 ms.

Three different pixels as a function of t_E were also included with their respective fittings, to show the quality of the data. Although the images were performed with the pellet surrounded by water, the 2D map presents only those pixels which have fitted the data with a relatively good quality. As the surrounding water presents T_2 and T_1 much longer than the liquid inside the pellet, due to the short range of either t_E or t_R covered in the experiments, the error in the fittings is larger, permitting a filtering by means of a numerical mask.

94   Chapter 4. H$_2$O$_2$ Reaction Evolution Studied by NMR Imaging

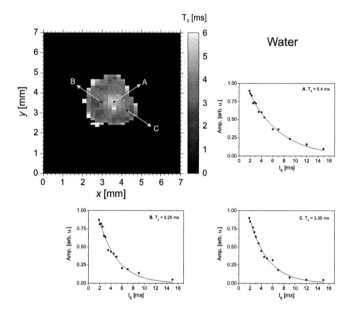

Figure 4.6: T_2−map of water inside a pellet whose metal has been removed from the surface. The parameters used for the image are: $FOV = 7 \times 7$ mm, 64×32 points in read and phase direction respectively, slice thickness 500 μm, NS=32 and $t_R = 1$ s. Three pixels as a function of t_E with the corresponding fittings are included.

Figure 4.7 shows the T_2−map obtained from the pellet saturated with 5 % H$_2$O$_2$ concentration. The same imaging parameters and t_E's as for the case of water were employed. Three pixels placed approximately in the same regions of the pellet as in the previous case are shown with the fittings, in order to compare. All the values in the map correspond to $T_2 < 3.05$ ms.

In Fig. 4.8 both maps are presented with the same T_2 greyscale, and a corresponding averaged profile. The profiles were obtained by adding 20 profiles in x−direction (20 profiles represent a high percentage of the whole pellet).

It is clearly observed that T_2 inside the pellet is larger for water compared to H$_2$O$_2$. Notice that while for the pellets used with water the relaxation time values close to the center are larger, decreasing as moving to the borders, for the pellet saturated

4.3. Reaction Evolution Inside the Pd-Catalyst   95

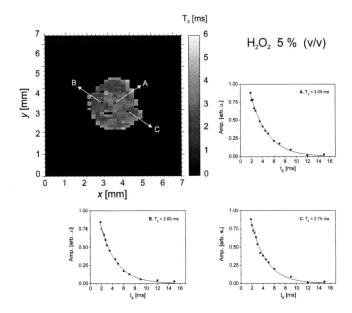

Figure 4.7: T_2–map of 5 % H_2O_2 inside the pellet without the metallic shell. The same parameters than in Fig. 4.6 were used.

with H_2O_2 a more uniform T_2 distribution is observed. It might indicate that both pellets have different porosity or pore size distribution. The bright pixels at the edge of the pellet surrounded by water are explained by considering contributions of liquid from outside and inside the sample.

It is important to remark that, although the calculations were made with the function shown in eqn. (4.27), in both cases shown above the B value was too small to be fitted (and then was set to be zero), except for a few pixels (less than 10) inside the pellet. The fact that the factor including the diffusion effect is too small is directly related with the short relaxation times: the signal dephases due to relaxation before any diffusion effect is possible to observe.

In order to obtain the so called T_1-maps, a series of images with different t_R was acquired, setting $t_E = 1.85\ ms$, i.e. the shortest technically possible with our device

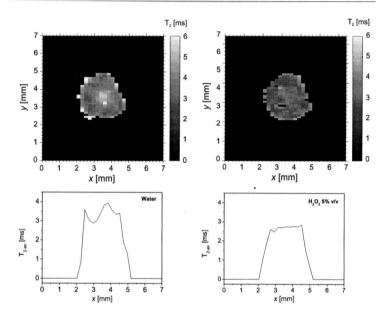

Figure 4.8: T_2–maps and averaged profiles for either water and H_2O_2 inside pellets containing no metal. The averaged profiles were obtained by adding 20 profiles in x–direction.

in order to maximize the signal-to-noise ratio. Keeping t_E constant, the third factor in eqn. (4.26) and any diffusion effect become a constant, and the resulting signal depends on T_1 through the factor $[1 - \exp(-t_R/T_1)]$. Thus, fitting I_p in the third dimension with a function of the form

$$I_p = I_0 \left[1 - \exp(-t_R/T_1)\right] \qquad (4.28)$$

a T_1–map is obtained.

Figure 4.9 shows the T_1–map of water inside the same pellet as that presented in Fig. 4.6. The same three pixels are plotted vs. t_R with the respective fittings. All the parameters other than t_E and t_R were set to be the same than in the T_2–maps experiments. The repetition time, t_R, was varied from image to image to cover the range $25 - 1000$ ms. Notice that, as in the case of T_2, T_1 values are larger close to the center of the pellet.

4.3. Reaction Evolution Inside the Pd-Catalyst

Figure 4.10 shows the T_1–map for 5 % H_2O_2 inside the same pellet used in Fig. 4.7, with the same three representative pixels and their respective fittings. Surprisingly, the values are slightly smaller than in case of water, whereas in bulk samples, T_1 is larger. The effect might be related with the extra oxygen on the molecules and its effect in the interaction of the spins with the paramagnetic porous walls.

Figure 4.11 shows both T_1–maps with averaged profiles, obtained adding the same 20 x–profiles than in the case of T_2. There, it can be observed once again a more homogeneous map in the case of the pellet saturated with H_2O_2. This homogeneity supports the assumption made above, about a difference in the porous properties of both pellets.

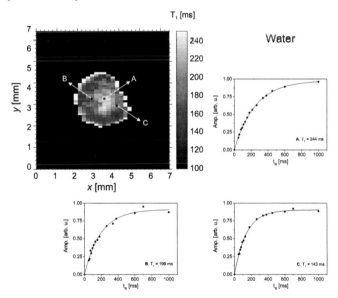

Figure 4.9: T_1–map of water inside the pellet without metal. The imaging parameters used here were set to be equal to those used in Fig. 4.6. The echo time was $t_E = 1.85$ ms while t_R was varied from 25 ms to 1 s.

Although it can be seen that T_2 as well as T_1 present smaller values when the pellet is saturated with H_2O_2 compared to being saturated with water, due to the

Chapter 4. H$_2$O$_2$ Reaction Evolution Studied by NMR Imaging

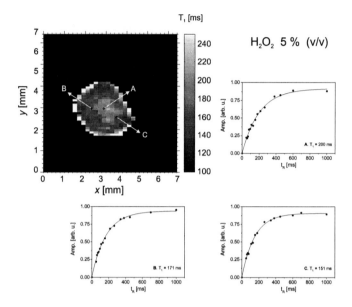

Figure 4.10: T_1–map of 5 % H$_2$O$_2$ inside the pellet without the metal-containing shell. The same parameters as in Fig. 4.9 were used.

apparent difference in porous properties it is not possible to reliably conclude that in one pellet, the values shown above will represent the final and initial values (respectively) of a decomposition.

In order to compare those values for the same pellet, the following experiment was performed:

(1) A pellet with the Pd shell was placed in the probe and once fixed, 1.1 ml of water was added.

(2) A T_2– and a T_1–map were acquired with the pellet being saturated with water.

(3) Then, the water was removed from the tube, and for more than 12 hours the pellet was allowed to dry.

(4) As a next step, 1.1 ml of H$_2$O$_2$ at 5 % v/v concentration was then dropped into the tube and the reaction started.

(5) The liquid was allowed to react for two hours until it has been considered

4.3. Reaction Evolution Inside the Pd-Catalyst

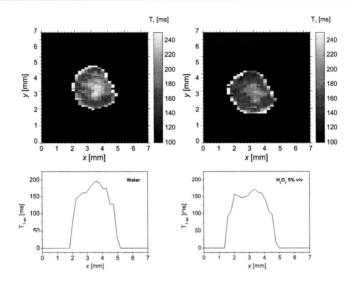

Figure 4.11: T_1–maps and averaged profiles for either water and H_2O_2 inside pellets containing no metal. The averaged profiles were obtained by adding the same 20 profiles as in Fig. 4.11.

that the H_2O_2 had completely penetrated.

(6) A T_2– and a T_1–map were acquired with the reaction going on, performing a CPMG experiment between every image in order to monitor the liquid evolution in the pellet vicinity.

In all the maps acquired during the decomposition, the T_2 outside have not changed more than 10 ms. As the reaction needed much more that 50 hours to cease, and T_2 outside had to go from 20 ms to 2.5 s (see Fig. 3.16), the maps can be considered to be acquired in almost static conditions (it represents less than 1 percent of the total change). The only extra effect might arise from the expected increase in the effective diffusion coefficient inside the pellet during the reaction. In order to estimate such an increase in the liquid transport a test experiment was performed. First of all, a pellet pre-saturated with D_2O was placed into the magnet. Then 1.1 ml of water was added and a series of images were acquired to see the water ingress into the pellet. The experiment was repeated with another pellet, also saturated

100   Chapter 4. H$_2$O$_2$ Reaction Evolution Studied by NMR Imaging

with D$_2$O, but 5 % H$_2$O$_2$ was added instead of water. With both sets of images it is possible to obtain an estimation of the time that the liquid front takes to reach, for instance, the center of the pellet. This provides a rough idea of the value of the effective diffusion coefficient inside the pellet. While in the case of water, that time was about 55 minutes, during the reaction experiment the value was about 45 minutes. Those values give for the effective diffusion coefficient a value between $3-5\times 10^{-10}\,m^2/s$, consistent with measurement of water diffusion coefficient inside the pellet by means of the PGSTE pulse sequence. Thus, no pronounced effect is expected in the pixel intensity due to diffusion in case of reaction.

Figure 4.12 shows T_2−maps of water (corresponding to item (2) above) and H$_2$O$_2$ reacting (item (6)) with average profiles, obtained by adding 20 x−direction profiles. The fittings had in general the same quality as those performed in previous experiments, and T_2 values show a similar difference, i.e. a difference of about 1 ms is expected between the beginning and the end of the reaction. Notice that for this pellet, no significant difference between the center and the rest of the volume is observed.

In Fig. 4.13 the result of the T_1−maps experiments are presented along with the averaged profiles. The values corresponding to the experiments with water are smaller compared to those obtained with the free-metal pellets. In agreement with the T_2−map shown in Fig. 4.12 no pronounced difference between the core and the borders is visible. Close to the metal shell larger values were encountered. These pixels correspond to liquid outside the pellet, but close enough to the metal layer to present a smaller relaxation time, permitting the fitting function to give an acceptable result. In the case of the liquid reacting, that effect is not as significant as previously, due to the motion induced by the rising bubbles. In average, T_1 in case of H$_2$O$_2$ reacting is slightly smaller than in the case of water, confirming the results obtained above.

4.3.3   The Optimum Contrast

In a decomposition starting with 5 % H$_2$O$_2$, as seen above, the T_2 is expected to increase inside the pellet, leading to an increase of the pixel intensity through the factor $\exp(-t_E/T_2)$, if t_E is kept constant. According to the previous discussion, any diffusion effect is neglected. On the other hand, T_1 will also increase during the reaction, and for a constant t_R will lead to a decrease in the pixel intensity by the

4.3. Reaction Evolution Inside the Pd-Catalyst

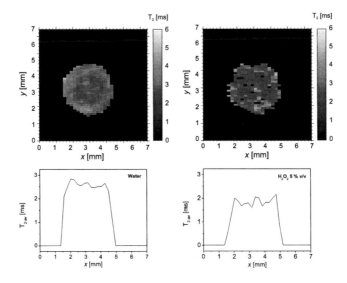

Figure 4.12: T_2−maps for water and H_2O_2 at the beginning of the decomposition. Averaged profiles are shown as well, obtained by adding 20 x−direction profiles.

factor $[1 - \exp(-t_R/T_1)]$. However, it is possible to find a combination of t_E and t_R to get the maximum change in the pixel intensity between both beginning and end of the reaction.

Although theoretically increasing the echo time will lead to a greater contrast between T_2 values at the end compared to the beginning of the reaction, it will be paid by sensitivity. Thus, we have set here the echo time to be the minimum possible, i.e. 1.85 ms, renouncing to a larger contrast in favor of larger signal-to-noise ratio. Taking as a reference the values from the last section, that echo time is close to the minimum T_2 expected, so going a bit further would represent a large loss of signal quality. In case of choosing $t_E = 2.85$ ms would result in a theoretical difference between end and beginning of the reaction of about 70 % instead of the 45 % obtained by setting $t_E = 1.85$ ms. But the loss of signal (more precisely of signal-to-noise) would be significant.

As the T_1 contrast is concerned, the safest choice would be to set t_R long enough to be larger than five times the longest T_1 expected. It would be translated into

102   Chapter 4. H₂O₂ Reaction Evolution Studied by NMR Imaging

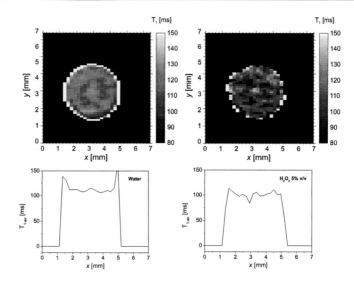

Figure 4.13: T_1–maps for water and H_2O_2 at the beginning of a decomposition. Averaged profiles obtained by adding 20 x–direction profiles are also included in order to allow comparisons.

a pixel intensity free of T_1 effect, i.e. the pixel at the end will have the same intensity than at the beginning of the reaction, when only T_1 is taken into account. Nevertheless, it will be paid with experimental time, and therefore with temporal resolution in the decomposition monitoring. Taking the values presented above as reference, t_R must be set to be 1 s, leading to a experimental time of about 17 minutes, assuming NS=32. On the other hand, it is known that for a given T_1 value, the optimum repetition time, where the best signal-to-noise ratio is obtained for a given experimental time, must be $t_R = 1.25\ T_1$ [Cal]. Assuming $T_1 = 100\ ms$ and setting $t_R = 125\ ms$, the difference between pixel intensity at the end compared to the beginning would be of about 5 %, while the experimental time is reduced to about 2 minutes. It represents a big improvement in time resolution, paid with a small loss in pixel intensity difference.

Considering all aspects mentioned above, assuming the pixel intensity at the beginning of the reaction is 3.5×10^{-4} in some units, repeating every 125 ms with an echo time of 1.85 ms, at the end one expects a pixel intensity of about 4.7×10^{-4}. This represents a change of 35 % (the choice of these numbers will become clear in what follows). Moreover, the major part of the change corresponds to the change in T_2, which in principle could be correlated with the changes in the liquid outside of the pellet, leading to the possibility of a monitoring similar to that described in the previous chapter.

4.3.4 Following a Reaction

We will present here results obtained by monitoring the pixel intensity inside the pellet during the decomposition of 5 % initial H_2O_2 concentration, for several hours. In all the reactions monitoring the same imaging parameters as in the relaxation maps have been used. According to the discussion above, t_R was set to be 125 ms and $t_E = 1.85$ ms for all the following experiments.

Figure 4.14 shows an image from a collection of 1400 images performed during a decomposition of 1.1 ml of liquid, for a period of about 50 hours. On the right hand side the intensity of a single pixel is shown vs. reaction time to illustrate the evolution. Although the increase in intensity is evident, with the values according to those calculated in the previous section, the curve is noisy. However, it is possible to smooth it by adding to every pixel intensity at a given time t, the values of a certain number of *neighbors* (neighbors in time domain). Denoting by n the number of points to be added, the intensity of the $i - th$ pixel at tiem t results

$$\bar{I}_p^i(t) = \frac{1}{n} \sum_{j=-n/2}^{j=n/2-1} I_p^i(t + j\Delta t) \tag{4.29}$$

where Δt is the time between images. The number n must be carefully chosen in order to not *erase* any fine detail in the curve. In the plot shown in Fig. 4.14, the black curve was obtained setting $n = 30$. Notice the intensity difference between the core and the border for this pellet.

Figure 4.15 shows a 2D image with 4 concentric circumferences, separating the image into excluding regions. The circumferences were labelled as C_i with $i = 1, .., 4$. The size of the circles were chosen in order to identify the pixels with approximately the same intensity (thus, relaxation times).

The regions were labelled as R_i with $i = 1, .., 4$, and can be described as,

- R_1 includes all the pixels placed inside the C_1.

104   Chapter 4. H₂O₂ Reaction Evolution Studied by NMR Imaging

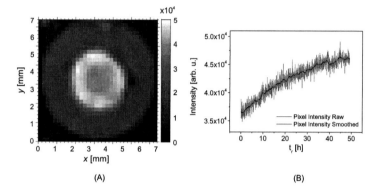

(A)   (B)

Figure 4.14: A 2D image extracted from a set of 1400 images acquired during the monitoring of one single reaction. The evolution of one particular pixel's intensity during the decomposition is plotted superimposed to the curve after the smoothing process to illustrate.

- R_i includes all the pixels placed between C_i and C_{i-1}, for $i = 2, 3$.

- R_4 includes all the pixels which intersect the circumference C_4.

As mentioned above, the experiment consisted of a series of 1400 images, with a CPMG experiment performed every 30 images, in order to follow the T_2 evolution of the liquid outside the pellet. The corresponding plot is shown in Fig. 4.15, where it is possible to observe that the 50 hours of reaction represent a large fraction of the total time needed to fully decompose the H_2O_2.

The bottom-left plot shows the mean pixel intensity vs. reaction time, averaged out within the different regions defined as above described. Although by adding the pixels the relative noise in the curves decreases, they were smoothed as well, with $n = 30$. The bottom-right plot shows the same curves, but shifted to coincide at the beginning of the reaction, in order to better compare them. Notice that the curve corresponding to R_4 presents the smallest change during the whole reaction time. It is expected due to the fact that, by using such short echo time, the $T_{2-\text{outside}}$ evolution only produces observable changes in the pixel intensity ($\propto \exp(-t_E/T_2)$) for small values of T_2, i.e. during the first few hours. The other three curves present a monotonic increase, with different expected limiting values (for $t_r \to \infty$) according

4.3. Reaction Evolution Inside the Pd-Catalyst 105

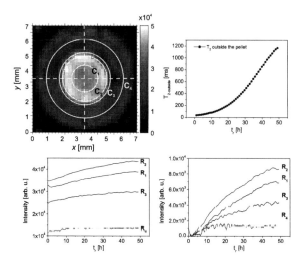

Figure 4.15: Results of monitoring an H_2O_2 decomposition in the presence of the Pd pellet. The 2D image represents 1 of the 1400 images acquired during the experiment. The circles were included to separate the pellet into different regions (see the text). The top-right plot shows the evolution of T_2 of the liquid outside the pellet. The plots in the bottom show the time evolution of the mean pixel intensity, averaged out in every region (left) and the same curves shifted to coincide in the origin, to allow clearer comparisons.

to the relaxation times within the respective regions.

Although the shape of the curves depends on the functional dependence of ρ, T_1 and T_2 on the reaction time t_r, a tentative exponential fitting was performed in all three curves, yielding approximately the same characteristic time, about 36 hours. It means that, despite the local differences in the porous distribution leading to differences in the relaxation parameters, the reaction evolves quasi-homogeneously within the pellet.

Several particles were tested later, in order to find one without presenting any appreciable difference in the images (i.e. pronounced differences in brightness). One pellet with similar size than that presented in Fig. 4.15 was found, and a new decomposition was monitored. In this case, however, the amount of liquid was halved, although the initial H_2O_2 concentration was kept to be 5 %, and the images

106   Chapter 4. H_2O_2 Reaction Evolution Studied by NMR Imaging

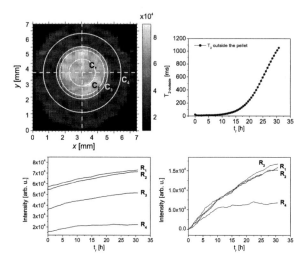

Figure 4.16: Results of monitoring an H_2O_2 decomposition in the presence of the Pd pellet during more than 30 hours. The data is organized in similar manner than in Fig. 4.15.

were acquired setting NS=64. A set of 520 images were performed during more than 30 hours of reaction, performing a CPMG experiment every 30 images, like before. Figure 4.16 shows the results, organized as in Fig. 4.15. Notice that, although no intensity differences can be observed here from the 2D image, the regions were defined as before, to directly compare. The evolution of $T_{2-outside}$ is slightly different compared to that observed in Fig. 4.15, mainly due to the unequal amount of liquid used.

The curves corresponding to the regions R_1 and R_2 do not show the pronounced difference observed before, due to the fact that no relaxation differences are present between the parts of the pellet. Shifting them to coincide at $t_r = 0$ it is possible to observe that they follow the same trend during the whole reaction, except, of course, the curve corresponding to R_4, as expected.

We can conclude that, it is possible to follow the decomposition inside the pellet by means of monitoring the pixel intensity, no matter whether differences in the pore space are present. Such differences will affect the final value in every region,

4.3. Reaction Evolution Inside the Pd-Catalyst

but the evolution itself presents the same features.

4.3.5 An Open Problem

The discussion and results presented above are valuable to prove that a simultaneous monitoring of a reaction in the interior as well as in the liquid surrounding the Pd pellet is possible, by combining NMR Imaging and NMR Relaxometry. The results show a relatively homogeneous evolution inside the porous medium, due to the fact that only effects averaged out over a long time are possible to observe. However, the situation would become less trivial if more particles are included, and/or the liquid is forced to flow through an array of catalyst particles. Here we will present a simple approach to the situation of a single pellet surrounded by other catalyst particles, in order to illustrate the problem. The new features encountered in these experiments represent an open problem and extra experiments and variants must be employed to arrive to a deeper comprehension. What follows is the statement of a problem rather than an attempt to explain the effect observed.

Figure 4.17 shows the situation expected in a real reactor schematically. In the zoom the contact points between a representative pellet and its neighbors are marked. In such situations, the transport of reactant to the metal layer will be obstructed, resulting in a reduced effective catalyst surface.

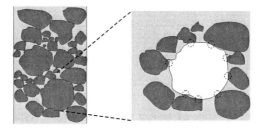

Figure 4.17: An array of pellets of different sizes in a reactor, shown schematically. The zoom at the right hand side shows a pellet and the contact points with the neighbors.

The intention was to simulate such situation in a controlled manner, in order to be able to increase the difficulty step by step. Thus, as a very first step it has been

108   Chapter 4. H_2O_2 Reaction Evolution Studied by NMR Imaging

opted to have a single pellet, reproducing the setup used until here, but modifying the surface by partially removing the Pd layer.

The pellet was fixed to the tube, with a tiny glass sphere used as a marker of the metal-free region. The tube was filled with water and 2D images were acquired in order to label the position of the modified surface. The position was carefully marked before removing the setup from the magnet, and the glass sphere. Thus, images with the pellet surrounded by water were acquired varying the imaging parameters in order to observe some extra contrast. No further contrast source was identified. Finally the same parameters used in previous section were adopted.

In Fig. 4.18, a 2D image of the modified pellet with a schematic representation is presented. In the NMR image, the position of the surface without metal is highlighted.

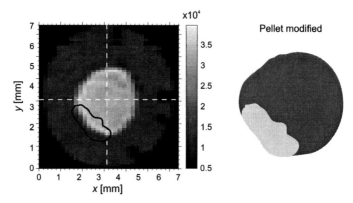

Figure 4.18: A 2D image of a Pd pellet with the metal surface partially removed. The region without metal is highlighted in the image. On the right hand side, the pellet modified is shown schematically.

T_2- and T_1-maps were acquired too, in case of the pellet submerged in water as well as with the reaction going on, as discussed in 4.3.2 (data not shown). The values, in terms of T_2 and T_1 differences between the end and beginning of the reaction, are in the same range as in the case of non-modified pellets.

One reaction was then monitored, like in the previous experiments. The total amount of liquid was further decreased to be 0.37 ml, in order to reduce the experimental time. A set of 850 images were acquired, with the same parameters as those

4.3. Reaction Evolution Inside the Pd-Catalyst 109

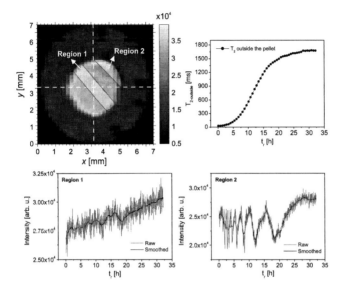

Figure 4.19: Results obtained by monitoring an H_2O_2 decomposition for 32 hours in the presence of a modified Pd pellet. The initial concentration was kept to be 5 % and the total amount of liquid was 0.37 ml. The image shows two regions defined according to the symmetry of the problem. The two plots at the bottom show the evolution of the average pixel intensity for both regions

used in previous experiments, performing a CPMG experiment every 30 images as well.

Figure 4.19 shows the results of the monitoring. From the evolution of $T_{2-outside}$ it can be seen that the reaction was close to the end after 32 h. The 2D image shows two regions defined according to the symmetry of the problem. The two plots at the bottom show the pixel intensity evolution for both regions, with the smoothing achieved by adding 30 neighbors in time. Notice that, while the pixels in Region 1 present the expected behavior, those pixels in Region 2 show a pronounced oscillation. This feature is exclusively observed for this sample, so it is reasonable to assume that the modifications made in the surface are responsible for it. During the processing time, many different variants for the definition of the regions were attempted, resulting always in curves as shown in Figure 4.19.

As the pixels presenting the oscillation cover a wider range of intensity values compared to the pixels showing just a monotonically increasing tendency, a new variable ($Fluct$) was defined as the difference between maximum and minimum pixel intensity. Figure 4.20 shows the $Fluct$−map. Notice that the maximum values correspond to the region of the surface without metal, and the geometrically opposed region, with metal. The other parts of the map have much lower intensity. It is important to remark that, $Fluct$−maps in the case of non-modified pellets were constructed, none of them presenting such a feature. On the right hand side, a plot of the averaged pixel intensities corresponding to both regions is presented. The symmetry of the oscillations is outstanding.

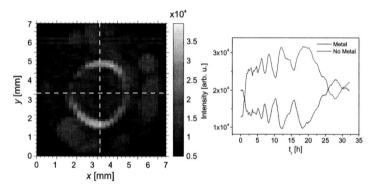

Figure 4.20: the $Fluct$−map clearly showing the regions where the pixel oscillations are present. The plot in the right hand side shows the evolution of the pixel intensity in both regions. The metal-free region is in the same position as shown in Fig. 4.18.

Figure 4.21A shows the same plot presented in Fig. 4.20 with the maxima and minima indicated by dotted lines. Notice the coincidence between minima in one curve and maxima in the other one, and vice versa.

The curves present an increasing mean value superimposed onto an oscillation. Notice that, as the reaction evolves and consequently the fluid motion outside and inside becomes slower, the frequency of the pixel oscillations progressively decreases (the separation between successive dashed lines increases in Fig. 4.21A). In order to better compare them, the mean value was subtracted from both curves, resulting in the plots shown in Fig. 4.21B. The fact that both curves are in counter phase

4.3. Reaction Evolution Inside the Pd-Catalyst 111

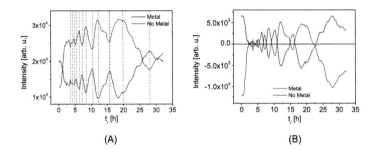

Figure 4.21: (A) Average pixel intensity corresponding to pixels placed in the surface where the metal removal was performed, and in the diametrical opposite region, both presenting the highest intensities in the $Fluct$-map shown in Fig. 4.20. The maxima and minima are marked by dashed lines. (B) the same plot after removing the mean value of both curves, in order to allow better comparisons.

becomes clearer.

Such an effect was first observed in the pellet presenting the modified surface. As shortly mentioned above, all the other experiments were checked afterwards in order to identify oscillation too, but no similar behavior was found. The experiment was repeated with another pellet, modified in the same manner, and placed in different position with respect to the gradient system in order to discard artifacts. The data, not shown here, presented exactly the same feature. So far, no satisfactory explanation was found for this effect. The only possible conclusion to be made at this stage is that the modification produced in the surface is directly related with the appearance of the oscillations. Furthermore, they are spatially linked as well, due to the fact that the pixels showing the oscillatory change in the intensity are placed in the region free of metal and the region diametrically opposed. All other possible conjectures are really difficult to hold, due to the nature of the oscillations: they keep in counter-phase during more than 30 hours. It is very risky to assume some collective motion of the liquid with such a long coherence on time, in a system producing bubbles and turbulence in the vicinity of the catalyst. This problem represents a challenge and underlines the importance of the non-invasivity of NMR not only in solving problems but also in identifying new problems to be solved.

Chapter 5

Conclusions and Outlook

Ideas are like rabbits. You get a couple and learn how to handle them, and pretty soon you have a dozen.

John Steinbeck

In the previous chapters, different experiments concerning the decomposition of aqueous hydrogen peroxide solutions catalyzed by the presence of porous media of different geometries and metal content were described in detail. Although the theoretical aspects, the experimental details and partial conclusions were provided in every single chapter, we will present here a brief summary of the more relevant results obtained along this work, accompanied in every case by a description of variants and/or extensions of the experiments that can be implemented relatively easily in the future. For the sake of clarity, the summary will respect the structure of the thesis, dealing with the results chapter by chapter, as they have been exposed.

Even though the 1-D average propagators acquired at different reaction times and spatial dimensions show the evolution of the reaction, the smaller magnetic field gradient strength required and the save in experimental time makes the effective diffusion coefficients a more suitable parameter for monitoring a decomposition, despite the fact that the information is more clearly extracted from its reaction time dependence. By comparing the relative shape of the curves it is possible to obtain qualitative and quantitative information about the reaction itself, allowing comparisons between catalyst particles with different: porosity, metal content, type of metal, etc. It is also possible to estimate the correlation time of the liquid moving around the pellet driven by the oxygen bubbles as a product of the reaction, at any time, by comparing the results of using either a PGSTE or a Double PGSTE pulse sequence, with the same total observation time.

If desired, the implementation of a slice selection can be achieved by simply transforming one, two or the three hard 90° pulses in Fig. 2.7A (or the firsts one, two or three pulses in Fig. 2.7B) into soft pulses, in order to measure the effective diffusion coefficient in any particular region.

This sort of experiments can be straightforwardly extended to the case of reactors including a net flow of the reactants. There are two mean strategies to remove effects from the extra decay produced by a coherent velocity distribution. One option is to first perform an experiment (or a series of experiments) without reaction, and consider the result as an offset to be subtracted from the results obtained with the reaction going on. Such an experiment could be made either by using another liquid possessing similar viscosity, or by replacing the catalyst particle with a free-of-metal pellet with the same dimensions. The other option could be to filter the much more coherent motion from the flow by means using a Double PGSTE with Δ carefully chosen. Both alternatives present advantages and disadvantages, and might be evaluated by the experimenter.

The experiments shown here as well as their alternative including flow in the setup, can be performed in standard desktop spectrometers, too. The values for the magnetic field gradient strength or the pulse sequence timing are not restrictive in those case, although the loss off sensitivity would lead to the necessity of more scans to be acquired per single experiment, directly translated into a reduction in reaction time resolution. Nevertheless, in experiments of about 18 hours, the temporal resolution obtained here (of about 2.5 *minutes*) is more than enough, and a decrease of that value in favor of the tremendous improvement in comfortability by using portable magnets is acceptable.

It is important to remark that, although all the diffusion experiments included here were performed in the catalytic decomposition of H_2O_2 , similar experiments can be performed in any other reaction involving a liquid-gas phase, due to the fact that no particular property of hydrogen peroxide was involved either in the data acquisition or in the post-processing.

Concerning to the use of proton exchange between H_2O_2 and water to monitor the reaction, the most relevant result, without any doubt, is the possibility of obtaining a reliable concentration quantification of H_2O_2 ranging from 5 % v/v down to 0.001-0.005 % v/v depending on the pH value, during a reaction, with a temporal resolution of 30 s. Nevertheless, the method presented here, although necessary to fully understand the process, is not the best to be implemented in future reactions monitoring. As an alternative, a much simpler setup can be achieved by buffering the H_2O_2 solutions to a fixed pH value. Thus, by measuring T_2, a concentration vs t_r curve can be obtained from the calibration curves. Notice that, if few pH values are going to be explored only, there is no necessity of repeating the entire 2D experiment shown in Fig. 3.16. T_2 vs C curves can be measured as a calibration with the buffered solution, resulting in a much shorter experiment.

Unfortunately, such experiments are not possible to extend to desktop devices as straightforward as in the case of the diffusion. As shown in Fig. 3.22, different behaviors are observed when decreasing the Larmor frequency, depending on the pH value, resulting in different range of concentrations where the technique can give acceptable results. If special care is taken in the confection of the calibration curves, reliable concentration quantifications can be obtained in the proper range.

The experiments can be performed in different solutions containing H_2O_2 too, with the only condition that the other liquids to be added to the aqueous hydrogen peroxide solutions do not produce decomposition. Thus, different bulk solutions can be prepared as before, and the corresponding calibration curves can be obtained.

The feasibility of extending these kind of experiments to reactors of continuous flow presents the limitations imposed by the following equation:

$$\frac{1}{T_{2-eff}} = \frac{1}{T_2} + \frac{1}{\tau_v}$$

where T_2 and T_{2-eff} are the transverse relaxation times obtained in a static and dynamic situation respectively, and τ_v the residence time of the liquid inside the sensitive volume of the coil (which is determined by the volume of the detection cell and the flow rate) [CSW]. In Fig. 5.1A simulated T_{2-eff} vs T_2 curves are presented, for different velocities. The sensitive volume was supposed to be cylindrical, with 15 mm length, and for simplicity the same velocity was supposed in the whole region.

The diagonal line shows the static situation. Up to $T_2 = 500\,ms$, the difference between T_{2-eff} and T_2 is below 5 % for $v = 1\,mm/s$ whereas for $v = 10\,mm/s$ even for $T_2 = 100\,ms$ the difference between the relaxation times is larger than 7 %. Fig. 5.1B shows the same plot only in the range $10 - 50\,ms$. If the liquid flows with $v = 30\,mm/s$, a $T_2 = 50\,ms$ will be affected about 5 %. In other words, the net flow of the reactants will tremendously limit the C range to be accurately measured.

However, the inconvenience can be avoided by allowing the flowing liquid to

116   Chapter 5. Conclusions and Outlook

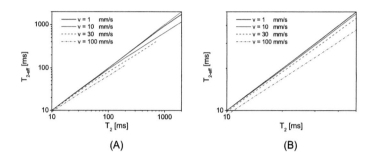

Figure 5.1: Effect of the flow in the transverse relaxation time. The T_{2-eff} values were obtained by simulating a cylindrical coil sensitive volume, with 15 mm length, for different liquid velocities. The velocity was considered to be constant through the the volume, for simplicity. (A) Relaxation times ranging from 10 to 2000 ms. (B) Only the range $10-50\,ms$ is shown to better observe the details.

react, stopping the flow periodically and performing a T_2 experiment in the static situation. It represent a disadvantage compared to the on-line monitoring, but for large amount of liquid reacting (and then long time reactions) this is an acceptable solution.

The experiments with NMR Imaging proved to be an alternative for the reaction monitoring, allowing to follow a decomposition inside a single catalytic particle, with the only condition of using pellets where the metal is placed near the outer surface. Although the temporal resolution is not sufficient to observe fine details during the reaction, a tendency is observed in a long time scale, with a spatial resolution of tenths of millimeters.

All what is shown in chapter 4 represents a first step, where the ideas are developed and all the limitations are explored, in order to prove the feasibility of being implemented in more complex situations. At this stage the results are promising, and with some variants the technique can be better exploited in further experiments. For instance, due to the porosity and tortuosity of the porous medium (mainly evidenced in the low diffusion coefficient within the pellet) no extra effects are expected if net flow is imposed to the reactant. The extension of the experiments shown here to the case of flow-reactors appears then to be straightforward.

On the other hand, the experiments performed with the surface-modified catalysts, as a simulation of a common situation in bed-pack reactors presented new features awaiting to be explained. The presence of the oscillations, highly spatial correlated for such a long time, represents an exciting problem and some variants can be tried. In the opinion of the author, one of the first experiments to be conducted in order to find more clues to arrive to a close explanation would consist in placing many pellets of different sizes with their surface entirely free of metal, surrounding the central pellet to be studied, with its surface untouched. In such a setup, only the pellet under investigation will react, but the geometrical effects produced by the numerous neighbors could be observed in more detail. As a more advanced variant, flow might be also added.

The future possible experiments can be improved by buffering the solutions, in order to maintain the pH constant during the whole decomposition time. It will be still impossible to transform the information obtained from the inner part of the pellet into H_2O_2 concentrations, due to the action of the porous walls. However, eliminating the pH variable, T_2 will be the only remaining parameter related to C, and in cases of globally homogeneous porosity as shown in Fig. 4.16 it will represent a direct indicator of the reaction progress.

To conclude this work, it is worth to remark, once more, the flexibility of NMR to be implemented in optimizing reaction processes. We believe that the results as well as the further experiments proposed here will help in the penetration of NMR in the field of on-line catalytic reactions monitoring, either in small scale reactors or in industrial sizes reactors. The latter application relies on the rapid development of movable NMR spectrometers with NMR Imaging capabilities [DMP+].

Bibliography

[ABBS] A. Amar, L. Buljubasich, B. Bluemich, and S. Stapf. *Transport Properties in Small-Scale Reaction Units*, chapter 16, pages 245–267. Wiley-VCH, 2008.

[Abr] A. Abragam. *Principles of Nuclear Magnetism*. Oxford University Press, 1983.

[AG1] A. Allerhand and H. S. Gutowsky. Spin-Echo NMR Studies of Chemical Exchange. I. Some General Aspects. *The Journal of Chemical Physics*, 41:2115–2124, 1964.

[AG2] A. Allerhand and H. S. Gutowsky. Spin-Echo Studies of Chemical Exchange. II. Closed Formulas for Two-Sites. *The Journal of Chemical Physics*, 42:1599, 1965.

[Akk] M. Akke. NMR Methods for Characterizing Microsecond to Millisecond Dynamics in Recognition and Catalysis. *Current Opinion in Structural Biology*, 12:642–647, 2002.

[ALM] M. Anbar, A. Loewenstein, and S. Meiboom. Kinetics of Hydrogen Exchange between Hydrogen Peroxide and Water Studied by Proton Magnetic Resonance. *Journal of the American Chemical Society*, 80:2630–2634, 1958.

[BDJ+] B. Bluemich, L. B. Datsevich, A. Jess, T. Oehmichen, Xiaohong Ren, and S. Stapf. Chaos in Catalyst Pores. Can We Use it for Process Development? *Chemical Engineering Journal*, 134:35–44, 2007.

[BKZ] M. A. Bernstein, K. F. King, and X. J. Zhou. *Handbook of MRI Pulse Sequences*. Elsevier Academic Press, 2004.

[Blu] B. Bluemich. *NMR Imaging of Materials*. Clarendon Press, Oxford, 2000.

[BP]   E. M. Bloembergen and R. V. Purcell. Relaxation Effects in Nuclear Magnetic Resonance Absorption. *Physical Review*, 73:679–712, 1948.

[BR]   P. S. Belton and R. G. Ratcliffe. NMR and Compartmentation in Biological Tissues. *Progress in Nuclear Magnetic Resonance Spectroscopy*, 17:241–279, 1985.

[Cal]   P. Callaghan. *Principles of Nuclear Magnetic Resonance Microscopy*. Clarendon Press, Oxford, 1991.

[CCS]   P.T. Callaghan, S.L. Codd, and J.D. Seymour. Spatial Coherence Phenomena Arising from Translational Spin Motion in Gradient Spin Echo Experiments. *Concepts In Magnetic Resonance*, 11:181–202, 1999.

[CF]   A. Caprihan and E. Fukushima. Flow Measurements by NMR. *Physics Reports-Review Section of Physics Letters*, 198:195–235, 1990.

[CHL+] S. Chou, H.Y. Huang, S.N. Lee, G.H. Huang, and C. Huang. Treatment of High Strength Hexamine-Containing Wastewater by Electro-Fenton Method. *Water Research*, 33:751–759, 1999.

[CP]   H. Y. Carr and E. M. Purcell. Effects Of Diffusion On Free Precession In Nuclear Magnetic Resonance Experiments. *Physical Review*, 94(3):630–638, 1954.

[CR]   J. P. Carver and R. E. Richards. A General Two-Site Solution for the Chemical Exchange Produced Dependence of T_2 Upon the Carr-Purcell Pulse Separation. *Journal of Magnetic Resonance*, 6:89–105, 1972.

[Cra]   J. Crank. *The Mathematics of Diffusion*. Oxford University Press, Oxford, 1975.

[CSW]   L. Ciobanu, J. V. Sweedler, and A. G. Webb. *Microcoil NMR for Reaction Monitoring*, chapter 2, pages 123–139. Wiley-VCH, 2006.

[Dat1] L. B. Datsevich. Alternating Motion of Liquid in Catalyst Pores in Liquid/Liquid-gas Reaction with Heat or Gas Production. *Catalysis Today*, 79:341–348, 2003.

[Dat2]   L. B. Datsevich. Oscillations in Pores of a Catalyst Particle in Exothermic Liquid (Liquid-Gas) Reactions. Analysis of Heat Processes and their Influence on Chemical Conversion, Mass and Heat Transfer. *Applied Catalysis A: General*, 250:125–141, 2003.

[Dat3]   L. B. Datsevich. Some Theoretical Aspects of Catalyst Behaviour in a Catalyst Particle at Liquid (Liquid-Gas) Reactions with Gas Production: Oscillation Motion in the Catalyst Pores. *Applied Catalysis A: General*, 247:101–111, 2003.

[Dat4]   L. B. Datsevich. Oscillation Theory Part 4. Some Dynamic Peculiarities of Motion in Catalyst Pores. *Applied Catalysis A: General*, 294:22–33, 2005.

[dG]   P. G. de Gennes. Theory of Spin Echoes in a Turbulent Fluid. *Physics Letters A*, 29:20–21, 1969.

[DMP+]   E. Danieli, J. Mauler, J Perlo, B. Bluemich, and F. Casanova. Mobile Sensor for High Resolution NMR Spectroscopy and Imaging. *Journal of Magnetic Resonance*, 198:80–87, 2009.

[EG]   H. Erlenmeyer and H. Gartner. *Helv. Chim. Acta*, 17:970, 1934.

[FAK]   I. Foley, Farooqui S. A., and R. L. Kleinberg. Effect of Paramagnetic Ions on NMR Relaxation of Fluids at Solid Surfaces. *Journal of Magnetic Resonance-Series A*, 123:95–104, 1996.

[Fuk]   E. Fukushima. Nuclear Magnetic Resonance as a Tool to Study Flow. *Annual Review of Fluid Mechanics*, 31:95–123, 1999.

[GH]   H. S. Gutowsky and C. H. Holm. Rate Processes and Nuclear Magnetic Resonance Spectra. II. Hindered Internal Rotation of Amides. *The Journal of Chemical Physics*, 25:1228–1234, 1956.

[GLM]   Ernest Grunwald, A. Loewenstein, and S. Meiboom. Rates and Mechanisms of Protolysis of Methylammonium Ion in Aqueous Solution Studied by Proton Magnetic Resonance. *The Journal of Chemical Physics*, 27:630–640, 1957.

[Gre] *Proceedings of the 3rd International Conference on Green Propellant for Space Propulsion and 9th International Hydrogen Peroxide Propulsion Conference, Poitiers, France*, 2006.

[GVW] H. S. Gutowsky, R. L. Vold, and E. J. Wells. Theory of Chemical Exchange Effects in Magnetic Resonance. *The Journal of Chemical Physics*, 43:4107–4125, 1965.

[Hah] E. L. Hahn. Spin Echoes. *Physical Review*, 80(4):580–594, 1950.

[HK] C.E. Huckaba and F.G. Keyes. The Accuracy of Estimation of Hydrogen Peroxide by Potassium Permanganate Titration. *Journal of the American Chemical Society*, 70:1604–1644, 1948.

[IT] R. Ishima and D. A. Torchia. Estimating the Time Scale of Chemical Exchange of Proteins from Measurements of Transverse Relaxation Rates in Solution. *Journal of Biomolecular NMR*, 14:369–372, 1999.

[JC] C.W. Jones and J.H. Clark. *Applications of Hydrogen Peroxide and Derivatives*. The Royal Society of Chemistry., 1999.

[Jen1] J. Jen. Chemical Exchange and NMR-T_2 Relaxation. *Andvances in Molecular Relaxation Processes*, 6:171–183, 1974.

[Jen2] J. Jen. Chemical Exchange and NMR-T_2 Relaxation. The Multisite Case. *Journal of Magnetic Resonance*, 30:111–128, 1978.

[JM] A. Jerschow and N. Müller. Suppression of Convection Artifacts in Stimulated-Echo Diffusion Experiments. Double-Stimulated-Echo Experiments. *Journal of Magnetic Resonance*, 125:372–375, 1997.

[KEM] R. Kleinberg, Kenyon W. E., and P. P. Mitra. Mechanism of NMR Relaxation of Fluids in Rock. *Journal of Magnetic Resonance-Series A*, 108:206–214, 1994.

[KH] J. Kaerger and W. Heink. The Propagator Representation of Molecular-Transport in Microporous Crystallites. *Journal of Magnetic Resonance*, 51:1–7, 1983.

[KM] Jozef Kowalewski and Lena Mäler. *Nuclear Spin Relaxation in Liquids: Theory, Experiments and Applications*. Taylor & Francis, 2006.

[KP] R. R. Knispel and M. M. Pintar. Temperature Dependence of the Proton Exchange Time in Pure Water by NMR. *Chemical Physics Letters*, 32:238–240, 1975.

[Lau] P. C. Lauterbur. Image Formation by Induced Local Interactions - Examples Employing Nuclear Magnetic-Resonance. *NATURE*, 242:190–191, 1973.

[Lev1] O. Levenspiel. *Chemical Reaction Engineering*. Whiley, 1999.

[Lev2] Malcolm H. Levitt. *Spin Dynamics*. Wiley, 2001.

[LL] Zhi-Pei Liang and Paul C. Lauterbur. *Principles of Magnetic Resonance Imaging*. IEEE Press, 2000.

[LM1] A. Loewenstein and S. Meiboom. Rates and Mechanisms of Protolysis of Di- and Trimethylammonium Ions Studied by Proton Magnetic Resonance. *The Journal of Chemical Physics*, 27:1067–1071, 1957.

[LM2] Z. Luz and S. Meiboom. Nuclear Magnetic Resonance Study of the Protolysis of Trimethylammonium Ion in Aqueous Solution. Order of the Reaction with Respect to Solvent. *The Journal of Chemical Physics*, 39:366–370, 1963.

[Mar] R. J. Marks. *Introduction to Shannon Sampling and interpolation Theory*. Springer-Verlag, New York, 1991.

[McC] Harden M. McConnell. Reaction Rates by Numclear Magnetic Resonance. *The Journal of Chemical Physics*, 28:430, 1958.

[Mei] S. Meiboom. Nuclear Magnetic Resonance. *Annual Review of Physical Chemistry*, 14:335–358, 1963.

[MG1] P. Mansfield and P. K. Grannell. NMR Diffraction in Solids. *Journal of Physics C*, 6:L422–L426, 1973.

[MG2] S. Meiboom and D. Gill. Modified Spin-Echo Method For Measuring Nuclear Relaxation Times. *Review Of Scientific Instruments*, 29(8):688–691, 1958.

[MLK+] Oscar Millet, P. J. Loria, Christopher D. Kroenke, Miquel Pons, and Arthur G. III Palmer. The Static Magnetic Field Dependence of Chemical Exchange Linebroadening Defines the NMR Chemical Shift Time Scale. *Journal Of the American Chemical Society*, 122:2867–2877, 2000.

[NC] O. Nalcioglu and Z. H. Cho. Measurement of Bulk and Random Directional Velocity Fields by NMR Imaging. *IEEE Transactions on Medical Imaging*, MI-6:356–359, 1987.

[PBS] D.L. Pardieck, E.J. Bouwer, and A.T. Stone. Hydrogen Peroxide Use to Increase Oxidant Capacity for In Situ Bioremediation of Contaminated Soils and Acquifers: A Review. *J. Contam. Hydrol.*, 9:221–242, 1992.

[PKL] A. Palmer, C. D. Kroenke, and J. P. Loria. *Methods In Enzymology, Nuclear Magnetic Resonance of Biological Macromolecules, Part B*, chapter 10, pages 204–238. Academic Press, 2001.

[Pri] W. S. Price. Pulsed-Field Gradient Nuclear Magnetic Resonance as a Tool For Studying Translational Diffusion: Part1. Basic Theory. *Concepts in Magnetic Resonance*, 9:299–336, 1997.

[RSB] Xiaohong Ren, S. Stapf, and B. Bluemich. *Multiscale Approach to Catalyst Design*, chapter 3, pages 263–284. Wiley-VCH, 2006.

[SB] N. A. Stephenson and A. T. Bell. Quantitative Analysis of Hydrogen Peroxide by ^{1}H NMR Spectroscopy. *Analytical and Bioanalytical Chemistry*, 381:1289–1293, 2005.

[SBS+] J. Svoboda, M. Blaha, J. Sedlacek, J. Vohlidal, H. Balcar, I. Mav-Golez, and M. Zigon. New Approaches to the Synthesis of Pure Conjugated Polymers. *Acta Chimica Slovenica*, 53:407–416, 2006.

[SH] S. Stapf and Song-I Han. *Introduction*, chapter 1, pages 1–45. Wiley-VCH, 2006.

[Sob1] W. T. Sobol. A Complete Solution to the Model Describing Carr-Purcell and Carr-Purcell-Meiboom-Gill Experiments in a Two-Site Exchange System. *Magnetic Resonance in Medicine*, 21:2–9, 1991.

[Sob2]   W. T. Sobol. A Complete Solution to the Model Describing Carr-Purcell and Carr-Purcell-Meiboom-Gill Experiments in a Two-Site Exchange System. *Magnetic Resonance in Medicine*, 21:2–9, 1991.

[ST]   E. O. Stejskal and J. E. Tanner. Spin Diffusion Measurements - Spin Echoes In Presence Of A Time-Dependent Field Gradient. *Journal Of Chemical Physics*, 42(1):288–292, 1965.

[Sta]   S. Stapf. Fluid Dynamics and Magnetic Resonance-Multidimensional NMR Techniques for the Visualization of Flow Patterns. In *RWTH Aachen University, Habilitation*, 2003.

[Ste]   J. Stepisnik. Analysis of Nmr Self-Diffusion Measurements by a Density Matrix Calculation. *Physica B*, 104:350–364, 1981.

[VDS+]   V. Vacque, N. Dupuy, B. Sombret, J. P. Huvenne, and P. Legrand. In Situ Quantitative and Kinetic Study by Fourier Transform Raman Spectroscopy of Reaction Between Nitriles and Hydroperoxides. *Journal of Molecular Structure*, 410:555–558, 1997.

[VK]   N. G. Van Kampen. *Stochastic Processes in Physics and Chemistry*. North Holland, Amsterdam, 1981.

[VRJM]   H. Voraberger, V. Ribitsch, M. Janotta, and B. Mizaikoff. Application of Mid-Infrared Spectroscopy: Measuring Hydrogen Peroxide Concentrations in Bleaching Baths. *Applied Spectroscopy*, 57:574–579, 2003.

[Wen]   H. Wennerström. Nuclear Magnetic Relaxation Induced by Chemical Exchange. *Molecular Physics*, 24:69–80, 1972.

[Woe1]   D. E. Woessner. Nuclear Transfer Effects in Nuclear Magnetic Resonance Pulse Experiments. *The Journal of Chemical Physics*, 35:41–48, 1961.

[Woe2]   D. E. Woessner. Brownian Motion and its Effects in NMR Chemical Exchange and Relaxation in Liquids. *Concepts in Magnetic Resonance*, 8:397–421, 1996.

[Woe3]   D. E. Woessner. *Encyclopedia of Magnetic Resonance*, chapter Relaxation Effects of Chemical Exchange, pages 4018–4028. Wiley, 1996.

[ZGA]   J. H. Zhong, J. C. Gore, and I. M. Armitage. Relative Contributions Of Chemical-Exchange And Other Relaxation Mechanisms In Protein Solutions And Tissues. *Magnetic Resonance in Medicine*, 11:195–308, 1989.

Acknowledgements

I consider that this part of the thesis should be the longest one. However, it does not seem to be a common use. In order not to be pedantic, I will follow the established as standard, and only mention the people who was strictly related with my work during the three years of my PhD studies.

First of all, I would like to thank **Prof. Siegfried Stapf** (Sigi!) for the unique opportunity to do my work under his supervision, but also for the thousands of kilometers he has covered between Ilmenau and Aachen just to discuss results and future experiments. I must not forget, though, the almost infinite patience and optimism when submitting papers and previous to conference presentations. Non strictly related to the thesis, but not less important, I do not have words to thank for being directly responsible of the three unbelievable months I have spent in New Zealand.

I would like to thank **Prof. Bernhard Blümich** for the opportunity of being part of his group in Aachen, and for permanently transmitting the motivation for going further with the experiments and the results interpretation.

Thanks to **Federico Casanova**, **Juan Perlo**, and **Ernesto Danieli**, for the time spent discussing my problems. Although my topic was far away from their interests, they were always willing to discuss details and results, naturally, accompanied by mate. In that way I must also thank **Andrea Amar**, not only for the discussions but also for the invaluable support in the long experiments leading to the construction of the T_2−surface.

Thanks to **Mihai Voda**, **Jörg Mauler**, and **Maria Baias**, my office partners, for the good times and for providing always a good environment to work.

Acknowledgements

Thanks to **Eva Paciok**, **Quinxia Gong** and **Jürgen Kolz**, for insisting in teaching me the rudiments of chemistry. Your effort was not in vane. Nowadays thank to the discussions with you, I can more or less suspect something related with chemistry.

Thanks to **Thomas Oehmichen** and **Leonid Datsevich** for their hospitality whenever being in Bayreuth, and for providing catalysts samples, as well as a few concepts and ideas about catalysis.

I don't want to forget to thank **Carlos Mattea**, an important friend, who always supported me and spent some nights discussing about experiments. But I must say thank you very much for the support in one of the most stressful times I had during the PhD, in Boston.

Special thanks to someone very important in my career: **Paul Callaghan**. It was a great honor for me to be part of his laboratory for three months, in that place so close to the paradise, New Zealand. The experience was invaluable from the scientific as well as from the personal point of view. In that way I must mention **Mark Hunter**, **Meghan Halse**, **Bruno Medronho** and **Guillaume Madelein**, with whom I shared a lot of good times at Wellington and surroundings. But I wouldn't be just if I don't mention all the people working there, who have created a great friendly atmosphere.

To finish I must thank all the people who shared with me the group in Aachen. I have spent there unforgettable three years.

Curriculum Vitae

Personal Details

Name	Lisandro Buljubasich Gentiletti
Date of Birth	07. 01. 1976
Place of Birth	Firmat, Santa Fé, Argentina

Carrier History

1982 - 1988	Primary School, Villa Regina, Río Negro, Argentina
1989 - 1994	High school, Villa Regina, Río Negro, Argentina
1995 - 2006	Physics student at FaMAF, UNC, Córdoba, Argentina
28th July 2006	Degree in Physics
2006 - 2009	PhD student at the Institut für Technische und Makromolekulare Chemie (ITMC), RWTH Aachen, Germany
15th March 2010	Defense of the doctoral degree in Natural Sciences